Research, Development, and Implementation of Pedestrian Safety Facilities in the United Kingdom

PUBLICATION NO. FHWA-RD-99-089

DECEMBER 1999

U.S. Department of Transportation

Federal Highway Administration

Research, Development, and Technology
Turner-Fairbank Highway Research Center
6300 Georgetown Pike
McLean, VA 22101-2296

FOREWORD

Creating improved safety and access for pedestrians requires providing safe places for people to walk, as well as implementing traffic control and design measures which allow for safer street crossings. A study entitled "Evaluation of Pedestrian Facilities" involved evaluating various types of pedestrian facilities and traffic control devices, including pedestrian crossing signs, marked versus unmarked crosswalks, countdown pedestrian signals, illuminated pushbuttons, automatic pedestrian detectors, and traffic calming devices such as curb extensions and raised crosswalks. The study provided recommendations for adding sidewalks to new and existing streets and for using marked crosswalks for uncontrolled locations. The "Evaluation of Pedestrian Facilities" also included synthesis reports of both domestic and international pedestrian safety research. There are five international pedestrian safety synthesis reports; this document compiles the most relevant research from the United Kingdom.

This synthesis report should be of interest to State and local pedestrian and bicycle coordinators, transportation engineers, planners, and researchers involved in the safety and design of pedestrian facilities within the highway environment.

Michael F. Trentacoste
Director, Office of Safety
Research and Development

NOTICE

This document is disseminated under the sponsorship of the Department of Transportation in the interest of information exchange. The U.S. Government assumes no liability for its contents or use thereof. This report does not constitute a standard, specification, or regulation.

The U.S. Government does not endorse products or manufacturers. Trade and manufacturer's names appear in this report only because they are considered essential to the object of the document.

1. Report No. FHWA-RD-99-089	2. Government Accession No.	3. Recipient's Catalog No.

4. Title and Subtitle Research, Development, and Implementation of Pedestrian Safety Facilities in the United Kingdom	5. Report Date
	6. Performing Organization Code

7. Author(s) David G. Davies	8. Performing Organization Report No.

9. Performing Organization Name and Address David Davies Associates 6 Hillside Road Norwich NR7 0QG United Kingdom University of North Carolina Highway Safety Research Center 730 Airport Rd, CB #3430 Chapel Hill, NC 27599-3430	10. Work Unit No. (TRAIS)
	11. Contract or Grant No. DTFH61-92-C-00138

12. Sponsoring Agency Name and Address Federal Highway Administration Turner-Fairbanks Highway Research Center 6300 Georgetown Pike McLean, VA 22101-2296	13. Type of Report and Period Covered
	14. Sponsoring Agency Code

15. Supplementary Notes

16. Abstract

This report was one in a series of pedestrian safety synthesis reports prepared for the Federal Highway Administration (FHWA) to document pedestrian safety in other countries. Reports are also available for:

Canada (FHWA-RD-99-090)
Sweden (FHWA-RD-99-091)
Netherlands (FHWA-RD-99-092)
Australia (FHWA-RD-99-093)

This is a review of recent research on pedestrian safety carried out in the United Kingdom. A comprehensive list of references is provided. The report covers many types of pedestrian facilities, the UK pedestrian safety record, as well as some education and enforcement matters. The report cites an access document with adequate references to allow further investigation of specific areas, and some commentary on research and implementation.

The past 5 years have seen increased attention given to road safety issues in the UK. Developments of particular relevance to pedestrians include greater emphasis on reducing vehicle speeds in urban areas through physical, legal, and publicity measures: also development of Puffin crossings and new operating strategies such as MOVA. However, while specific facilities can affect safety at individual sites, improvements in overall safety for pedestrians requires a comprehensive road safety strategy that is fully integrated with land use and transport policy.

17. Key Words: pedestrians, pelican crossing, zebra crossing, puffin crossing, traffic calming, tactile pavement surfaces	18. Distribution Statement

19. Security Classif. (of this report) Unclassified	20. Security Classif. (of this page) Unclassified	21. No. of Pages 47	22. Price

Form DOT F 1700.7 (8-72)   Reproduction of form and completed page is authorized

SI* (MODERN METRIC) CONVERSION FACTORS

APPROXIMATE CONVERSIONS TO SI UNITS

Symbol	When You Know	Multiply by	To Find	Symbol
LENGTH				
in	inches	25.4	millimeters	mm
ft	feet	0.305	meters	m
yd	yards	0.914	meters	m
mi	miles	1.61	kilometers	km
AREA				
in²	square inches	645.2	square millimeters	mm²
ft²	square feet	0.093	square meters	m²
yd²	square yards	0.836	square meters	m²
ac	acres	0.405	hectares	ha
mi²	square miles	2.59	square kilometers	km²
VOLUME				
fl oz	fluid ounces	29.57	milliliters	mL
gal	gallons	3.785	liters	L
ft³	cubic feet	0.028	cubic meters	m³
yd³	cubic yards	0.765	cubic meters	m³

NOTE: Volumes greater than 1000 l shall be shown in m³.

Symbol	When You Know	Multiply by	To Find	Symbol
MASS				
oz	ounces	28.35	grams	g
lb	pounds	0.454	kilograms	kg
T	short tons (2000 lb)	0.907	megagrams (or "metric ton")	Mg (or "t")
TEMPERATURE				
°F	Fahrenheit temperature	5(F-32)/9 or (F-32)/1.8	Celcius temperature	°C
ILLUMINATION				
fc	foot-candles	10.76	lux	lx
fl	foot-Lamberts	3.426	candela/m²	cd/m²
FORCE and PRESSURE or STRESS				
lbf	poundforce	4.45	newtons	N
lbf/in²	poundforce per square inch	6.89	kilopascals	kPa

APPROXIMATE CONVERSIONS FROM SI UNITS

Symbol	When You Know	Multiply by	To Find	Symbol
LENGTH				
mm	millimeters	0.039	inches	in
m	meters	3.28	feet	ft
m	meters	1.09	yards	yd
km	kilometers	0.621	miles	mi
AREA				
mm²	square millimeters	0.0016	square inches	in²
m²	square meters	10.764	square feet	ft²
m²	square meters	1.195	square yards	yd²
ha	hectares	2.47	acres	ac
km²	square kilometers	0.386	square miles	mi²
VOLUME				
mL	milliliters	0.034	fluid ounces	fl oz
L	liters	0.264	gallons	gal
m³	cubic meters	35.71	cubic feet	ft³
m³	cubic meters	1.307	cubic yards	yd³
MASS				
g	grams	0.035	ounces	oz
kg	kilograms	2.202	pounds	lb
Mg (or "t")	megagrams (or "metric ton")	1.103	short tons (2000 lb)	T
TEMPERATURE				
°C	Celcius temperature	1.8C+32	Fahrenheit temperature	°F
ILLUMINATION				
lx	lux	0.0929	foot-candles	fc
cd/m²	candela/m²	0.2919	foot-Lamberts	fl
FORCE and PRESSURE or STRESS				
N	newtons	0.225	poundforce	lbf
kPa	kilopascals	0.145	poundforce per square inch	lbf/in²

*SI is the symbol for the International System of Units. Appropriate rounding should be made to comply with Section 4 of ASTM E380.

(Revised September 1993)

TABLE OF CONTENTS

TABLE OF CONTENTS (Con't)

Page

1. Introduction

1.1 Basis of report

The report has been compiled on the basis of the following:

- Literature search using Silverplatter CD-ROM data base held at the Transport Research Laboratory (TRL) library;
- Meeting of UK technical experts held at the Department of the Environment, Transport and the Regions (DETR);
- Consultation with various academics and practitioners in local government; and
- Review of relevant literature from a wide variety of sources, including literature search and material assembled over the past 5 years.

1.2 Purpose and scope of report

The aim of this report is to give an overview of the issues regarding research, development, and implementation of pedestrian facilities in the United Kingdom. It concentrates on the period 1993 to 1997.

The report covers many types of pedestrian facilities, the UK pedestrian safety record, as well as some education and enforcement matters. The report provides an access document with adequate references to allow further investigation of specific areas, and some commentary on the research and implementation.

The report concentrates on safety aspects of pedestrian facilities, rather than issues of pedestrian convenience or promoting walking as a mode of transport. There are some inevitable overlaps and even conflicts between pedestrian safety and convenience. These are addressed where necessary but the emphasis of the report is on pedestrian facilities and pedestrian safety.

Pedestrian safety has long been a concern of central government, local government, and others in the United Kingdom. Over the past few years, particularly the period of this study, it has received greater attention partly because of the growing importance attached to promoting walking for transport, environment, and health reasons, and also because of concerns that whereas the UK road safety record was generally good, it is less satisfactory for pedestrians.

2. Pedestrian Safety in the United Kingdom

2.1 Casualty statistics

Road accidents and casualties are the conventional means of assessing safety. In Great Britain[1] in 1996, there were 320,302 reported casualties of which 46,381 (14.5%) were pedestrians (DETR,1997a).

---

[1] The main source of published road accident data is *Road Accidents Great Britain*, published annually. Consequently, most of the accident data presented in this report are for Great Britain (not the United Kindgom) and exclude Northern Ireland. In 1995 the total number of road accidents in Northern Ireland was 11,725.

In certain areas and sub-groups pedestrians form a higher percentage of casualties. They make up approximately one in three of all road users killed, some 20 percent of casualties in built-up areas, and casualty rates for child pedestrians (particularly 11-13 years old) and elderly pedestrians are particularly high. Details of pedestrian accident statistics are provided annually in *Road Accidents Great Britain* (DETR, 1997a). Commentaries on the pedestrian statistics and trends can be found in the background paper the *National Walking Strategy* (DOT, 1996a), Lambert (1997) and, on child pedestrians, O'Reilly (1994).

The GB pedestrian fatality rate is around 2.0 per 100,000 population. In 1994, the GB (and the UK) rate was the eighth lowest when compared to other European Union member countries, and almost the same as the United States rate (2.2). Differences in exposure (the amount of walking) population profile, modal split, and other factors may explain many of the differences and need to be taken into account when making comparisons. The GB fatality rate (all modes) is 6.4 per 100,000 population; this is the lowest overall in Europe (Sweden, 6.7: Netherlands, 8.5; and Finland, 9.5) and considerably lower than some other countries (US, 15.6; Greece, 20.3; Portugal, 28.7) (DETR 1997a, Table 48).

There are significant regional differences in pedestrian casualty rates within the United Kingdom. Wales has the lowest rate (1.8 per 100,000) followed by England (2.0), Scotland (2.2), and Northern Ireland (2.7).

The number of GB pedestrian fatalities declined rapidly at the end of the second World War (1945) to 2063 in 1952, then rose during the 1960s to a peak of 3,153 in 1966, and then declined during the 1970s to 2,335 in 1976, with a steep decline from the late 1980s onwards. Each year since 1990 has seen a new record low number of pedestrian fatalities. Total pedestrian casualties have also declined but at a considerably lower rate from an average 1981-85 of 61,741 to 47,028 in 1996.

Calculating pedestrian casualty rates in relation to pedestrian exposure is difficult because adequate pedestrian activity data are generally lacking. A major study in the United Kingdom (Ward et al, 1994) did provide accident rates for a range of pedestrian environments and pedestrian groups. This found that, for example, there were 411 casualties per 100 million km (61 million mi) walked or 66 casualties per 100 million roads crossed.

Reported accidents — the "official"road accident statistics, recorded by the police on STATS 19 forms and collated by local and central government — are known to underestimate the severity of pedestrian injuries (Hopkin et al., 1993). Also, pedestrian accidents not involving a vehicle (falls, trips, etc.) are not classified as road accidents and are not reportable to the police. The number of footway falls by pedestrians is hard to determine accurately but some local authorities are now paying more for injury compensation claims than they are spending on footway maintenance (Kindred Associations, 1995). Accidents between pedestrians and cyclists are rarely reported, and some police forces instruct staff not to record them.

2.2   National road safety targets

Targets for 2000

In 1987, national road casualty reduction targets were set by the Government. A "headline" target of a one third reduction from the 1981-85 average was to be achieved by the year 2000. There were also subsidiary targets for certain groups. Targets for pedestrians were to:

- reduce deaths by 40 percent
- reduce serious injuries by 40 percent
- reduce all casualties by 35 percent

The targets were based on trends and projections, plus estimates of what might be achieved through concerted efforts. To date the target for fatal pedestrian casualties has been achieved (table 1 and figure 1 show casualties during recent years). The DETR publishes an annual review of road safety which includes progress towards targets (DETR, 1997b). The projection for the period 1997-2001 for all road user casualties is for a slight fall in fatal and serious injuries and a slight rise in slight injuries (DETR, 1997c p26). Falls in pedestrian casualties are likely to be outweighed by rises in casualties to drivers and passengers.

Targets beyond 2000

Road safety targets for beyond the year 2000 are now being devised (PACTS 1995). It has been suggested that there should be separate targets for fatal/serious injuries and slight injuries; and that the targets should relate to levels of exposure, particularly for pedestrians and cyclists. Some organizations have argued that measures of danger, such as vehicle speeds and modal

Table 1. UK Pedestrian and Car Casualties for 1981-1996

Year	Car Casualties	Pedestrian Casualties	Pedestrian Fatalities
1981-85	143942	61741	1863
86	159178	60875	1841
87	159468	57453	1703
88	170705	58843	1753
89	184688	60080	1706
90	190558	60230	1694
91	179357	53992	1496
92	185645	51587	1347
93	187457	48098	1241
94	195109	48653	1124
95	193992	47029	1038
96	205277	46381	997

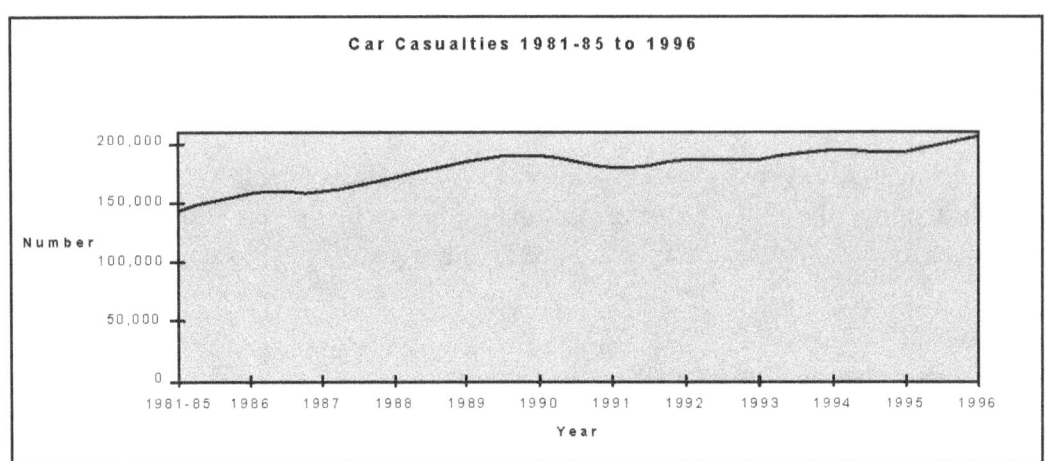

Figure 1A. Car occupant casualties.

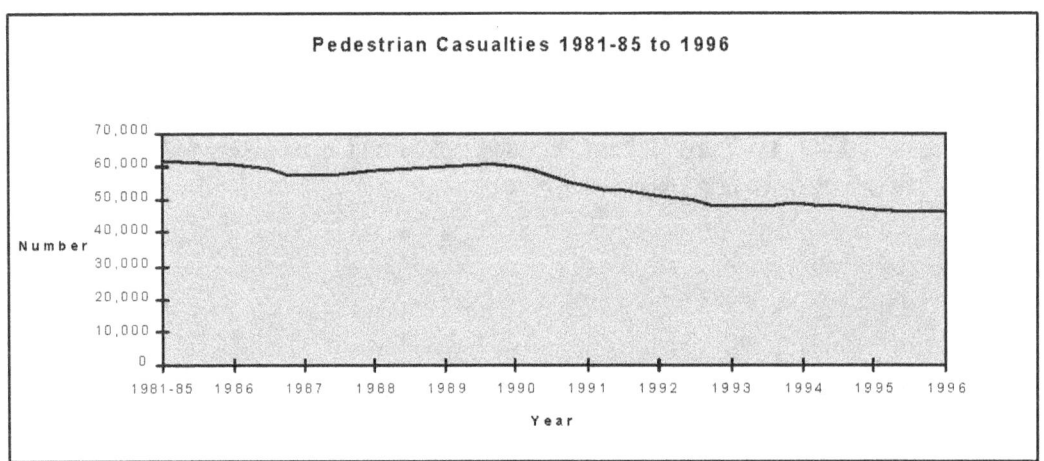

Figure 1B. Pedestrian casualties.

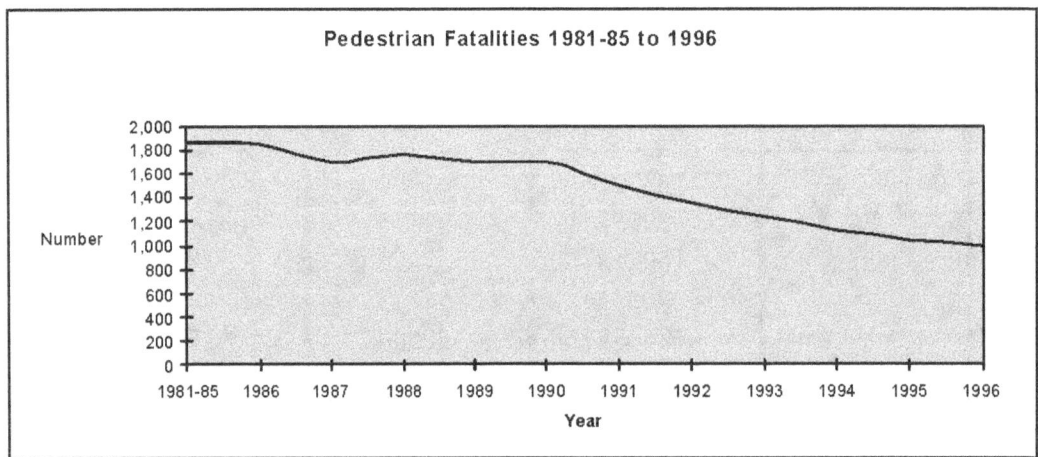

Figure 1C. Pedestrian fatalities.

Figure 1. Casualties and fatalities.

split, should be included in the targets in addition to casualty numbers. At this stage (early 1998), the new targets are, at least publicly, yet to be decided. However, the Government has announced that new targets will be adopted for the period 2000-2010 and that these will include a headline target and subsidiary targets for different severities (LTT 1997).

2.3   Explaining the trends

Describing changes in accident numbers and trends is relatively straightforward. It is far more difficult to say, with hard evidence, the reasons for the reduction in casualty numbers and equally difficult to say whether declining accidents signify improvements in road safety in a broader sense.

All road users

Commenting on the fall in all accident numbers compared to the 1981-85 baseline (figure 2), the DETR states:

"It is not possible to say with certainty why most casualty rates and numbers have reduced. There are many factors in play, but it is likely that the principal reasons are:

> **exposure** : people are walking, cycling and motorcycling less, so the number of casualties is falling. The number of car occupant slight injuries is rising at least partly because of the increase in traffic;
> **safer cars:** because cars are more robustly constructed and seat belt wearing rates have risen, car casualties are less severe;
> **safer roads** : new road construction and local safety measures (e.g., traffic calming) have contributed to preventing accidents and reducing casualty severity;
> **anti drink drive:** the number of fatal accidents in which an involved driver had been drinking over the limit has fallen by 57 percent since the baseline;
> **changing attitudes:** there is less tolerance of road accidents than there was. In 1987, when the target was set, there was deep concern about public acceptance of road casualties. " (DETR, 1997b p5)

Pedestrians

Over the period of 30 years, there has been a major shift from walk trips to car trips and this has accelerated over the past few years as car ownership has increased. The average distance walked declined by 18 percent between the 1981-85 baseline and 1995 (DETR 1997c) (figure 3). There has been a particularly large decline in walking by children, one of the pedestrian groups with the highest casualty rate. There has also been increased concern about personal security, and it is possible that this has reduced walk journeys after dark, which would also tend to have a disproportionate effect on accident numbers. Demographic changes also need to be considered. "It is not clear that the rate [per 100,000 km walked] of pedestrian casualties has fallen, and an upturn in serious injuries to child pedestrians since 1993 suggests that, for this subgroup, the risk is increasing." (DETR, 1997c p11)

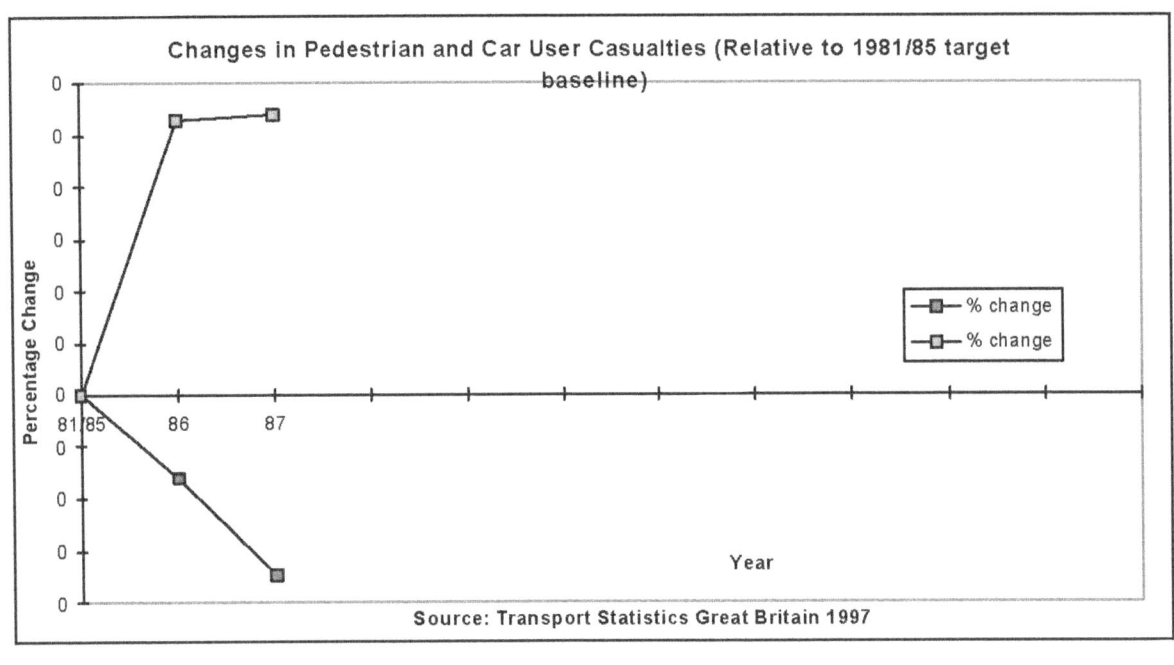

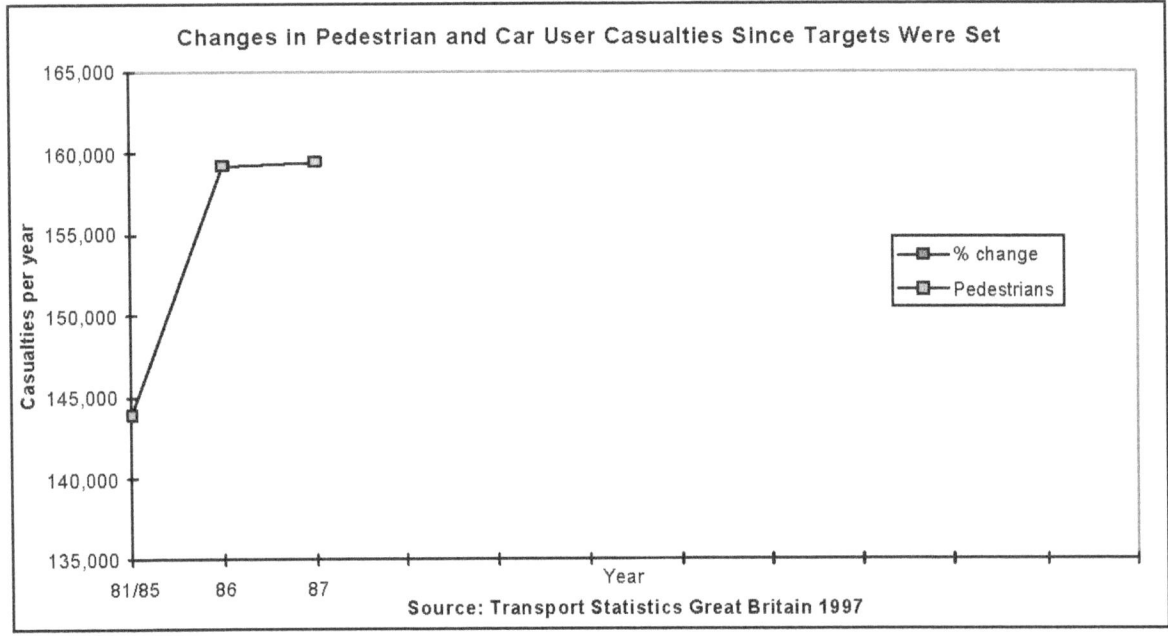

Figure 2. Changes in pedestrian and car user casualties.

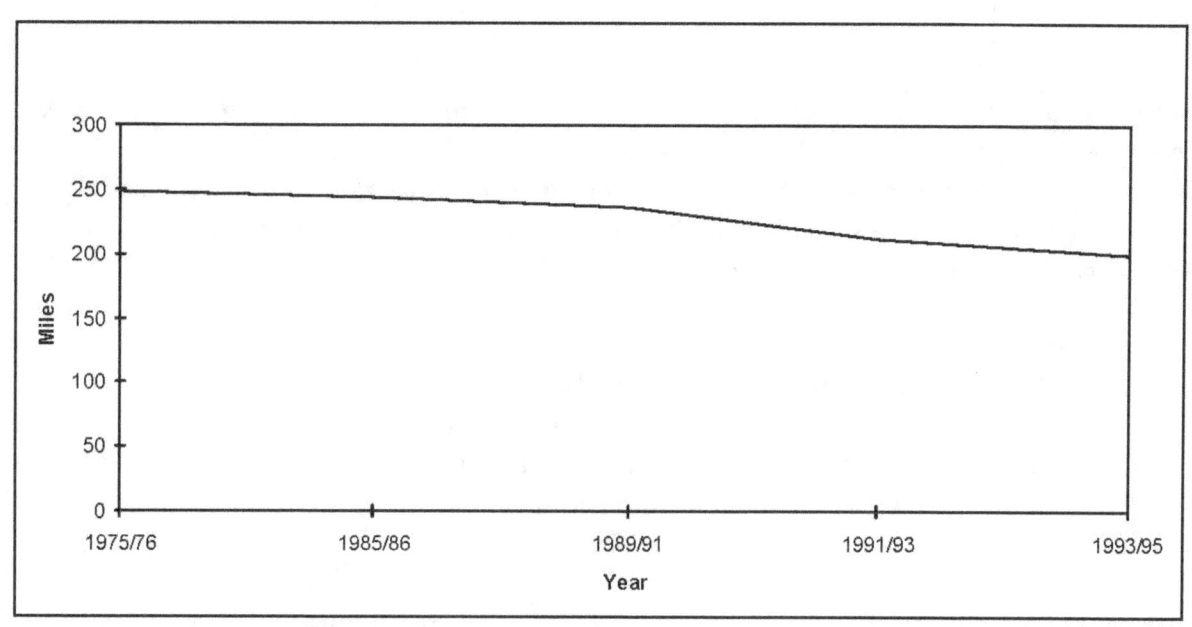

Figure 3. Average distance walked per person per year (1975-1995). (1 mile = 1.61 km)

Ninety-five percent of pedestrian casualties occur on "built-up" roads — defined as roads with a speed limit of 64 km/h (40 mi/h) or less. Most of these have a 48 km/h (30 mi/h) limit. Several research projects have attempted to address the fundamental question of why accidents occur, including pedestrian accidents. These include Carsten et al (1989) — all road users; Lawson (1990) — child pedestrians; and Davies and Winnett (1993) — all pedestrians.

Among the common factors identified by these studies were:
- The ordinariness of the circumstances leading to the accident.
- The drivers almost always considered the pedestrian to be at fault.
- Vehicle or road defects were rarely a significant cause or contributory factor.
- Consumption of alcohol by drivers and pedestrians.
- Masking of pedestrians by parked or stationary vehicles.

Conclusions that are drawn from these and other studies are that ordinary speeds and the 48 km/h (30 mi/h) speed limit are often too high for pedestrian safety and that it is unrealistic to expect child pedestrians to observe the same standards of traffic behaviour as adult road users. More fundamental assessments of society's propensity for accidents — inevitable results of its willingness to accept a certain amount of risk — are provided by Adams (1985).

2.4   Promoting walking and pedestrian safety

The issue of pedestrian safety has been given a new significance in the past 5 years because of increased concern about congestion and the environmental effects of traffic. Various influential reports have addressed this including that of the Royal Commission on Environmental Pollution (1994) which called for pedestrian fatality rates to be reduced from 2.2 to 1.5 by 2000. National land-use planning policy now requires that new developments are located so that they are accessible to pedestrians (DOE and DOT, 1994). The UK government, in cooperation with other agencies and voluntary bodies is currently drawing up a National Walking Strategy intended to halt and possibly reverse the decline in the amount of journeys walked, as part of an integrated transport policy. Some local authorities have already devised walking strategies for their areas. A comprehensive walking strategy has been drawn up for London (London Planning Advisory Council, 1996) although this does not deal in detail with the safety of pedestrian facilities (figures 4 and 5).

The decline in children walking to school is particularly relevant. Although few accidents occur on the journey to and from school, parents regularly cite traffic danger as the main reason for not allowing their children to walk (or cycle) to school. Perception of danger is difficult to measure and even harder to reliably compare over time or between places and individuals. However, it is probably true to say that the ordinary person in the street would not agreed with the professional's view that the UK's roads are now safer for pedestrians. There is some policy tension between targets to reduce road accidents and policies to encourage walking (and other sustainable but vulnerable modes such as cycling).

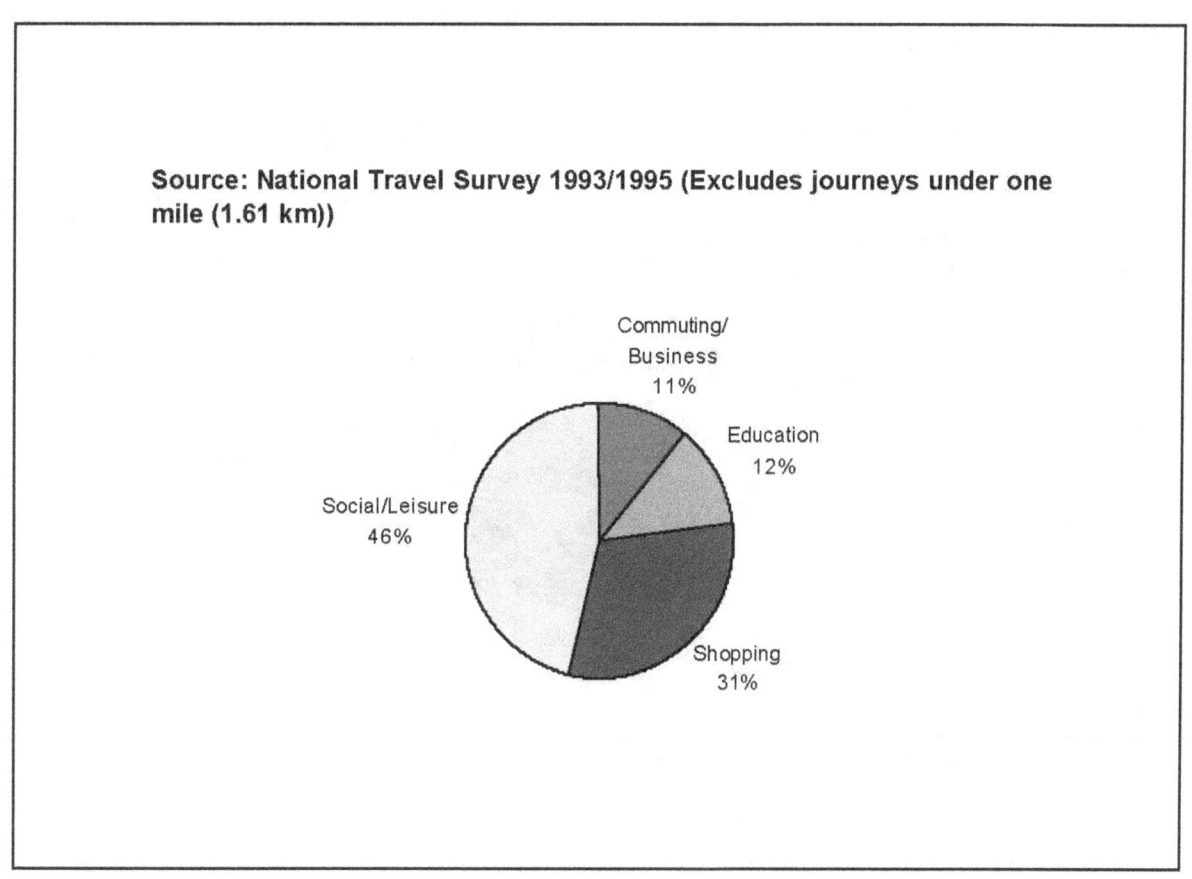

Figure 4. Distance walked per person per year by journey purpose.

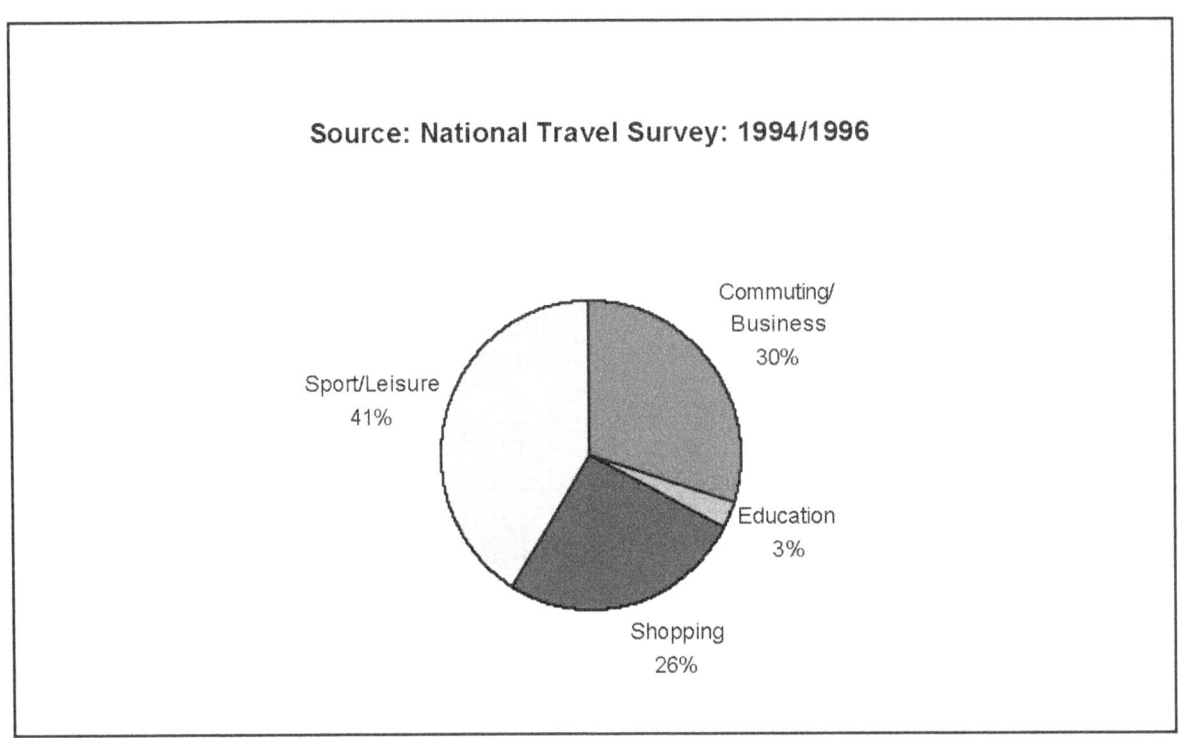

Figure 5. Distance travelled per person per year by car by journey purpose.

Although the United Kingdom probably has as much walking as most other comparable countries, and a lower pedestrian fatality rate than most, there is still some concern that more walking may lead to more casualties. "It would not be sensible to encourage people to walk more if this led to more casualties." (Lambert, 1997). Increased amounts of walking may lead to more pedestrian casualties but there are likely to be offsetting reductions, particularly if increased walking is at the expense of trips made by other modes and results in changes in journey lengths and patterns. The experience of cities such as York, which have achieved their casualty reduction targets while promoting walking and cycling, suggest that a simple comparison of casualty rates per 100,000 km is misleading (White, 1994).

This issue was addressed by the House of Commons Transport Committee (1996a) in its report on road safety for vulnerable road users. It concluded that both safety and use should be promoted and that there should be greater emphasis on reducing the dangers to pedestrians and cyclists and not restricting their movement. (For the Government's response, see House of Commons Transport Committee, 1996b.)

2.5   Child and elderly pedestrians

Whilst the UK pedestrian casualty rate of 2 per 100,000 pop is low, the rate for child pedestrians is acknowledged to be higher than in most other European countries: 1.3 fatalities per 100,000 children, nearly one third higher than the European Community average (Lambert 1997). The child pedestrian accident problem is addressed in DETR (1997c) and in section 9 of this report. A detailed study was undertaken by Lawson (1990). The causes of child pedestrian accidents are also analysed by Lynam and Harland (1992) and O'Reilly (1994). The higher rate of child pedestrian accidents amongst poorer families has also been recently investigated (Christie, 1995).
The DETR has recently revised its strategy for improving child pedestrian safety. In remarkably plain terms, it states that, from now on, more responsibility will be placed on drivers to avoid accidents with children (DOT, 1996b).

Elderly pedestrians are also acknowledged to be at a higher risk than average; these over 60 years old make up only one fifth of the population yet half of all pedestrian fatalities.

2.6   Conclusion

Pedestrian casualties are declining but so is the amount of walking. Whether pedestrians perceive improvements in their safety is not clear. A common perspective seems to be that, as the roads get busier with motorized traffic, with higher speeds and acceleration, safety is reduced.

3.   Overview of Accident Countermeasures and Safety Programs

3.1   Policies and priorities

The DETR carried out a major review of road safety policy in the mid-1980s (DOT, 1987a). This led to the adoption of national road safety targets — notably the one third reduction in accidents by the year 2000 — and a long-term program of measures. As a result, local highway authorities are required to draw up a road safety plan and periodically update it. The DETR produces road safety reports (DOT, 1995a; DETR, 1997b and DETR, 1997c), outlining objectives, priorities, policy, research, and performance against targets. The more detailed document (DETR, 1997c) examines each road user group and relevant safety measures separately.

In its latest review (DETR, 1997b), the main problems are summarized as:

- Excessive and inappropriate speed.
- Drinking and driving.
- Novice drivers.
- Protecting vulnerable road users.
- Reducing slight injuries.

Topics relating to safety of pedestrians, that have received new or increased DETR attention over the past 5 years include the following:

- Speed reduction publicity campaigns.
- Traffic calming.
- 32 km/h (20 mi/h) zones.
- Speed enforcement cameras.
- Child pedestrian safety.
- New forms of signal-controlled pedestrian crossings.

Pedestrian safety issues that have been highlighted or implemented by other safety interests, such as local highway authorities or non-government organizations (NGOs) include:

- Lower speed limits.
- Increased driver responsibility.
- Safe routes to schools.
- Road danger reduction.
- Safety audit.
- Urban safety management.
- Traffic reduction.

The common denominator for most of the public sector road safety work over this period has been attention to speed. (The motor industry has concentrated its efforts on protecting occupants in increasingly high-performance vehicles (Mackay, 1994).) Speed limits are commonly exceeded. The national road traffic speed surveys regularly show that a majority of motor vehicles exceed speed limits. For example, 70 percent of cars, 70 percent of light vans (<3.5 tonnes gross weight) and at least 46 percent of heavy goods vehicles exceed the 48 km/h (30 mi/h) urban speed limit (DOT, 1996c).

Whilst speed is by no means the only causal factor in accidents (pedestrian or otherwise), excessive speed has been shown to be a contributory factor in a high percentage of accidents, and is likely to lead to more serious injuries if an accident occurs.

The booklet *Killing Speed, Saving Lives* (DOT, 1992a) summarized much of the research into speed undertaken by the TRL, outlined Government priorities for further research, and a series of measures to reduce excessive speed.

During the 1990s, emphasis has shifted from localized treatment of accident "blackspots" to area or route treatment, often involving traffic calming and speed reduction. This also reflects the Urban Safety Management approach set out in the Institution of Highways and Transportation's guidelines (IHT, 1990).

3.2   Evaluating accident countermeasures

As noted in section 2, pedestrian accidents have declined sharply over the past 5 years. However, establishing causes and effects is not easy. As noted earlier, the amount of pedestrian activity had also declined sharply. In addition, the evaluation of the accident reduction effects from specific countermeasures has rarely been rigorous, partly because of practical difficulties. The typical method consists of a comparison of reported accidents (or casualties) for 3 years before and after the scheme. It rarely includes the known confounding factors such as changes in traffic flow, changes in traffic composition (particularly pedestrian flows), background trends in accident numbers, regression to mean effects, adaptive behaviour by vulnerable road users or more controversial aspects such as accident migration. Elvik (1997), reviewing United Kingdom and other accident studies, found that very few allowed for these factors; he also found that when they were taken into account, little or no accident reduction benefit could be directly attributed to the countermeasure. Elvik's work supports the earlier work of Adams (1985) and others, suggesting a trade-off between safety and performance. To compound the problem of evaluation, no work appears to have been done to show how the (claimed) accident savings from particular schemes or program relate to the overall changes in accident numbers. Accident countermeasures can alter the relative convenience and the balance of risks between user groups, but it is more difficult to prove cause and effect or conclusive accident savings.

3.3   Road danger reduction

Although there is widespread agreement that road safety should be improved and casualties reduced, there are disagreements over policies and priorities. The conventional approach has been criticized for concentrating exclusively on reducing accident numbers (the bottom line), if necessary by removing or restricting vulnerable road users, while ignoring the dangers (fast moving motor vehicles) which cause the accidents. This has resulted in the establishment of the Road Danger Reduction Forum supported by several local authorities, such as Leeds and York, and environmental organizations (Davis, 1993). The Forum favors measures such as lower speed limits, greater legal responsibility for drivers towards vulnerable road users, and promotion of benign modes of transport (walking, cycling, and public transport). Some of these policies, such as promoting walking and cycling, are now also Government policies. However, the central thrust of the Government's road safety policy is casualty reduction, and it remains to be seen to what degree road danger reduction issues will be integrated with road safety strategy.

4. Overview of Pedestrian Facilities

4.1 Planning and design

Pedestrian facilities have typically been provided as part of road schemes, or in response to accident problems, or for amenity or economic reasons — usually in town centers. Pedestrians have rarely been catered for as traffic, i.e., their needs assessed and provided for in terms of journey origins, destinations, desire lines, flows, and levels of service.

There is no comprehensive manual on pedestrian facilities for the United Kingdom (although guidelines are currently being produced under the auspices of the IHT). Initial reference points are *Transport in the Urban Environment* (IHT, 1997), *Design Manual for Roads and Bridges* (DOT, 1995b) and *Design Bulletin 32 Layout of Roads in Residential Areas* (DOE and DOT, 1992).

All significant new highway schemes, including pedestrian facility schemes, are now supposed to be checked at appropriate design stages by an independent safety audit. The revised safety audit guidelines (IHT, 1996) place greater emphasis on the safety of vulnerable road users and recommend auditors mentally and physically walk the scheme to thoroughly consider it from a pedestrian safety perspective. Safety audit is now well established within the design procedures of UK highway authorities and, from a safety perspective, is considered very beneficial. However, the procedures are narrowly focused and, if applied overzealously, or in isolation, can over-ride broader objectives, such as pedestrian convenience and overall transport policy and urban design objectives.

4.2 Types of pedestrian crossings

The principle types of pedestrian crossings in the United Kingdom are as follows:

(a) Midblock Crossings Without Signal Control (Crosswalks)

> *Zebra crossing.* Indicated by black and white bands painted on the carriageway. Pedestrians on the crossing have priority over vehicles.

> *Pedestrian refuge island.* Consists of kerbing, bollards and signs in the center of the carriageway, enabling pedestrians to cross more easily, in two stages. No pedestrian priority.

> *Curb build-out.* Consists of curbing, bollards, and signs at the edge of the carriageway, reducing the crossing width and making pedestrians more visible to drivers. No pedestrian priority.

> *Flat-top road hump.* A hump usually 75 to 100 mm high designed to reduce vehicle speeds and to enable pedestrians to cross on the level (at grade). No pedestrian priority. The photographs that comprise figure 6 illustrate these features.

Figure 6. Examples of pedestrian facilities in the United Kingdom including zebra crossing, curb build-out, flat-top speed hump, and refuge island.

(b) Midblock Crossings With Signal Control

Pelican crossing. Pedestrian light controlled crossing. Activated by pedestrian pushing the button. A "red/green man" signal on far side of the carriageway shows pedestrian when to cross.

Puffin crossing. Pedestrian User-Friendly intelligent crossing. Also activated by the pedestrian pushing a button. Intended as a replacement for the Pelican, it monitors the presence of pedestrians waiting and crossing and lengthens or shortens the crossing time accordingly. The "red/ green man" signal is located on the near side to the pedestrian.

Toucan crossing. "Two can cross." Similar to the Pelican and the Puffin but shared with bicycles. (See figures 7 and 8.)

(c) Pedestrian Phase at Traffic Signals

A "red/ green man" signal on far side of the carriageway shows pedestrian when to cross. Activated by pedestrian pushing the button.

Further details on the above crossings are given in the sections that follow.

4.3   Assessment framework for pedestrian crossings

Official guidance on whether a pedestrian crossing should be provided and, if so, what sort of crossing is most suitable, is contained in Local Transport Note 1/95 (DOT, 1995c). This recommends use of an assessment framework. The site should be surveyed approximately 50 m either side of the proposed crossing point and all relevant information recorded, including:

- Carriageway and footway type and width.
- Surroundings.
- Vehicular/pedesrtian flow and composition.
- Average crossing time and dificulty of crossing.
- Road accidents.

The crossing options should then be assessed against the relevant factors which are likely to include:

- Difficulty in crossing.
- Peak hour vehicle delay.
- Carriageway capacity.
- Vehicle speeds.
- Local representations.
- Cost.

LTN 1/95 introduced a more comprehensive and flexible assessment procedure than was previously required (DOT, 1987b). It replaces the PV^2 criterion where P = pedestrian flow and

Figure7. Signal hardware and pressure sensitive mat at experimental Puffin Crossing.

Figure 8. Pedestrians using Puffin Crossing.

V = vehicle flow: the general rule was that a Pelican crossing should only be installed if $PV^2 > 1 \times 10^8$ (although other factors, such as proximity to a school or hospital, could be taken into account if the PV^2 criterion was not met). Although now officially superseded, PV^2 remains in day-to-day use and comparison of the methods is interesting.

The planning, design, and installation of pedestrian crossings are prescribed in Local Transport Note 2/95 (DOT, 1995d). This covers all types of at-grade crossing, including pedestrian refuges, Zebra crossings, and various types of signal-controlled crossings. Advice is given in relation to the proximity of junctions, school crossing patrols, visibility, crossing width, guard railing, crossing approach, surfaces, disabled pedestrians, lighting, signing, bus stops, and street furniture. Under the Road Traffic Regulation Act 1984, it is no longer necessary for local highway authorities to obtain approval from the Government for installation or removal of a pedestrian crossings. However, they should consult locally and inform the DETR.

4.4   Relative safety of crossings

It has been shown that providing a pedestrian crossing does not necessarily reduce pedestrian casualties, partly because the crossing may cause changes in levels and type of pedestrian activity. Similarly, it is not possible to say that one type of crossing is safer than another. "Each type has advantages and disadvantages; the type chosen should be appropriate to the circumstances of the site and the demands and behaviour of the road users." (DOT, 1995c)

5.   Pedestrian Crossings without Signal Control (Crosswalks)

5.1   Zebra crossings

Zebra crossings were introduced in their current design in the 1950s and are still in widespread use in the United Kingdom. They are indicated by black and white bands painted on the surface of the carriageway. Since 1971 zigzag lines have been painted upstream and downstream of Zebra crossings to alert drivers to the crossing and prohibit overtaking and parking close to the crossing. There may or may not be a pedestrian refuge at the mid point on the crossing.
(See figure 9.)

Drivers are required, under the Highway Code, to stop for pedestrians on Zebra crossings. Legally, pedestrians have to establish precedence by standing on the crossing. UK drivers often also stop when they see a pedestrian waiting to cross — something that does not appear to happen at Zebra crossings in other countries.

Over the past 10 years, many Zebra crossings have been replaced by Pelican crossings (signal control), and new crossings tend to be Pelicans rather than Zebras. There is now an estimated 9,000-10,000 Zebra crossings in the United Kingdom, down from 13,000 in 1981. The reason for this and the research basis are discussed below.

Figure 9. Typical zebra crossing in the United Kingdom.

"There is little difference in the average rate of personal injury accidents at Zebra and signal-controlled types. At individual sites, however, the type of crossing selected may have considerable effect on the future accident record." (DOT, 1995c).

Broadly speaking, Zebra crossings are considered inappropriate on high speed or high motor traffic flow roads, particularly multi-lane roads. The DETR guidance recommends that Zebras should not be installed on roads where 85 percentile speed is greater than 56.35 km/h (35 mi/h).

Since 1991 Zebra crossings can be raised, i.e. combined with a flat-top road hump, to produce a humped Zebra. This makes it easier for pedestrians with pushchairs, trolleys, wheelchairs, etc. to cross and helps to reduce motor vehicle speeds and emphasizes that drivers should give way to pedestrians crossing. However, the number of such crossings is still small.

Zebra crossings generally cause far less delay to pedestrians than Pelican crossings (Hunt, 1997). They are considerably cheaper to install and maintain than signal-controlled crossings. However, there has been a tendency for traffic engineers to replace Zebras with Pelicans and to choose Pelicans rather than Zebras when installing new pedestrian crossings. This is because of a number of factors. Firstly, as traffic flows have risen, there has been a considerable increase in the number of signal-controlled junctions and UTC systems. Signal-controlled pedestrian crossings (i.e., Pelicans) are seen as more appropriate in this environment, on the assumption that drivers are concentrating on signals to indicate stop rather than other visual cues. Secondly, where pedestrian demand is heavy, Pelican crossings allow motorized traffic to continue to flow. Thirdly, there is sometimes a perception, by both public and engineers, that Pelicans are safer and better than Zebras because drivers are controlled by signals rather than using their discretion.

When accidents have occurred at Zebra crossings, there has been a tendency to replace the Zebra with a Pelican in the hope that this will solve the problem. It certainly demonstrates that something has been done but it does not necessarily improve safety (or convenience) for pedestrians.

There are some signs that the Zebra may be making a come-back, particularly within traffic calmed areas, because Zebra crossings give pedestrians greater priority and are less visually intrusive and less expensive. Also, it is argued that Pelican crossings encourage the driver to look for signals and not for pedestrians and that this can have a detrimental effect on pedestrian safety.

Edinburgh has recently installed three Zebra crossings on arms of three busy roundabouts in the city center (George Street). Pedestrian flows are high. These are the first Zebra crossings to be installed in Edinburgh for 30 years. The reasons for installing Zebras were that they gave greater priority to pedestrians and reduced pedestrian delay; also that Pelican crossings would have been difficult to install in these locations. The performance of the Zebras, in terms of safety, delay, etc., will be monitored. Initial reports from officers indicate that they are working well and that, when motor vehicle queues build up, pedestrians stop to allow the vehicles to proceed, an interesting reversal of the usual priority.

York, another UK city known for promoting walking and pedestrian safety, has also begun to reverse its earlier policy of replacing Zebra crossings with Pelicans. This was being done to integrate pedestrian crossings with the UTC system, in order to reduce vehicle delay. Two humped Zebra crossings have recently been installed, partly because they give pedestrians greater priority and also because, being considerably cheaper than Pelican crossings, they can be installed more widely. Zebra crossings have also recently been installed in Norwich.

5.2   Pedestrian refuges

Since 1990 there has been a dramatic increase in the installation of refuge islands, particularly pedestrian refuges. (Local highway authorities already had powers under the Highways Acts to install refuges to assist pedestrians, but the 1991 Traffic Calming Regulations permitted them to also install refuges for non-pedestrian purposes.) This increase has been due to efforts to reduce pedestrian casualties and to assist pedestrians to cross roads on which traffic has grown substantially. Pedestrian refuges have also formed part of traffic calming schemes and local safety schemes designed to provide greater separation between oncoming vehicles and to reduce overtaking accidents. However, to a large extent, pedestrian refuges have been installed in preference to other traffic calming and road safety measures because they are easy to install. They do not require a Traffic Regulation Order, and there is no requirement to consult the public or others before installing. They require less signing and are generally less controversial or constrained by regulations than road humps. There has been a tendency, therefore, to install traffic islands and pedestrian refuges in response to perceived needs to do something.

Pedestrian refuges can provide a series of crossing points along a road where it would be impractical to install Zebras or Pelicans at each crossing location. The minimum recommended width (across the road) for a refuge is 1.2 m, but 2 m is preferred to accommodate wheelchairs, pushchairs, and cycles. Where pedestrian flows are high, this may need to be increased, although in such cases another form of crossing (Zebra or Pelican) may be more appropriate.

Detailed research into the effects of installing pedestrian refuges was undertaken by Thompson et al (1990) in Nottingham. A survey of 32 sites showed significant decreases ($p<0.1$) in 85 percentile speed at 9 sites and significant increases in speed at 4 sites (although all increases were at one scheme). Somewhat surprisingly no relationship was found between the residual width of the road (between 3 m and 4.5 m) and the proportion of vehicles exceeding the 48 km/h (30 mi/h) speed limit. Although the schemes were introduced to improve pedestrian safety, there was a slight increase in pedestrian accidents. This was presumably caused by an increase in pedestrian activity but this aspect was not measured nor were traffic flows. Regarding the change in accidents of all types, "a statistically significant reduction was only achieved at two of the schemes. In addition, the reduction in accidents at all of the schemes combined was not significant when compared to accident control data." Residents perceived that pedestrians' safety had improved (which is quite compatible with increased accident numbers if pedestrian activity increased) but perceived that safety for pedal cyclists and drivers was reduced.

Research into the effects of road narrowings, including pedestrian refuges, was carried out because of concerns about effects on pedal cyclist safety (Davies et al, 1997). This found that although there appeared to be a relationship between motor vehicle speed and carriageway width, other factors, such as traffic flows, pedal cycle flows, and congestion, were probably more significant than width in influencing speed.

Overall, it seems that pedestrian refuges assist pedestrians to cross roads more easily, with less delay and greater perceived safety. However, vehicle speeds are not necessarily reduced and pedestrian accidents may not reduce if pedestrian activity increases. There may also be adverse effects such as parking problems and problems for pedal cyclists. Unfortunately, as with so much road safety engineering research, the studies do not include exposure data so an overall assessment of the safety benefits is difficult to obtain. It may well be that pedestrian refuges do have greater safety benefits for pedestrians than the Nottingham accident data imply.

5.3   Curb build-outs

Curb build-outs may be used in isolation or in conjunction with other measures such as Zebra crossings or pedestrian refuges. They increase the prominence of pedestrians, particularly if there is curb-side parking, and reduce the crossing width. A study of an early scheme in Nottingham (Thompson and Heydon, 1991) where pedestrians were often masked by parked vehicles, found a reduction in average pedestrian accidents from 4.7 per year to 1 per year. The build-out extended 2.5 m into the carriageway and included substantial lengths of guard rail.

Most build-outs away from parked vehicles are considerably narrower, usually less than 1 m (3.28 ft) into the carriageway. As build-outs are often part of more comprehensive measures, specific evaluation of build-outs in isolation has been limited. As with pedestrian refuge islands, build-outs can cause concern to cyclists who are forced closer to motor vehicles.

5.4   Flat-top road hump

Many traffic calming schemes have been introduced in the United Kingdom in the past 5 years. These often include flat top road humps which make crossing more convenient for pedestrians and potentially safer by concentrating pedestrian crossing movements, particularly when parked cars may mask children crossing. They may be used in conjunction with curb build-outs. Flat-top road humps, like pedestrian refuge islands, give no precedence to pedestrians over vehicles. However, unlike the situation at refuges, drivers often give way to pedestrians as if they were approaching a Zebra crossing. This is because vehicles need to slow down for the road hump. It is more likely to happen where traffic is queuing and the number of pedestrians crossing is high. These schemes have been criticized for introducing ambiguity: some pedestrians behave as if they have some legal priority and while some drivers give way, by no means all do so. However, they are generally successful in that they provide pedestrians with safer crossing locations that are easier to use (particularly for those pedestrians with push-chairs (buggies), shopping trolleys, or wheelchairs) and reduce pedestrian delay.

Since the Road Humps Regulations 1996, local highway authorities in England have been permitted to install road humps without a prior speed reducing feature, although DETR generally advise against doing so.

6. Pedestrian Crossings with Signal Control

6.1 Pelican crossings

The main type of independent or stand-alone signal-controlled pedestrian crossing in the United Kingdom is the Pelican crossing which was introduced in 1969. In 1979 local authorities were permitted to install Pelicans without special authorization from the DOT. Major changes were introduced in the 1987 Pelican Crossing Regulations. (DOT, 1991). Since 1979, Pelican crossings have been widely introduced and estimated to be more than 11,000 Pelican crossings in the United Kingdom (Hunt and Lyons, 1997).

The Pelican crossing has a far-side red man/green man signal aspect. Pedestrians register a demand at the push-button and approaching traffic is monitored, usually by microwave detectors. Towards the end of the pedestrian crossing phase, the green man changes from constant to flashing. There is a three-aspect signal head to control vehicles, including a flashing amber phase which permits drivers to go if all pedestrians have cleared the crossing. For pedestrians, however, the red/green man is advisory only; there is no offense of jay walking, or equivalent, in the United Kingdom. Modern Pelican crossings are now Vehicle Actuated, i.e., they monitor approaching vehicles and recall the pedestrian phase more quickly if there are no vehicles approaching. However, substantial numbers of Fixed Time Pelican crossings remain in operation.

As noted earlier, the criteria for deciding if and what type of crossing should be provided is set out in LTN 1/95. The installation of a Pelican (or other type of crossing) will not necessarily reduce pedestrian accidents. It may even result in some increase in pedestrian accidents because of increased pedestrian activity or other factors. Studies have attempted to find relationships between accident rates and levels of pedestrian and vehicle flow. A recent study (CSS, 1997), however, found no correlation.

6.2 Puffin crossings

During the 1990s the DOT has been sponsoring experiments with other types of signal-controlled crossing, particularly the Puffin and the Toucan (below). The Puffin crossing is intended as a replacement for the Pelican.

The need to develop improved pedestrian crossings, and the research program, is described by Billings and Walsh (1991). The Puffin has been developed in response to shortcomings of the Pelican, namely

- Inadequate time for slow pedestrians to cross.
- Stressful and confusing nature of flashing green man.
- Unnecessary delay to vehicles when pedestrians push the button but are able to cross before the green man shows; also when pedestrians complete the crossing early.
- Excessive delays for pedestrians because of fixed minimum time between pedestrian phases.

At traffic signals with pedestrian phases (but not at Pelicans), there is also the "dead spot" between the green man being extinguished and the red man appearing. (There is no flashing green /flashing amber.) This can cause confusion and anxiety to pedestrians.

Research has been undertaken to develop new technical equipment, and operating strategies, and to assess user behavior. Davies (1992) provides the results of the first Puffin experiments, supported by European Community funds from the DRIVE program. These Puffins used pressure-sensitive mats at the curbside to detect waiting pedestrians and infra-red on-crossing detection to adjust the pedestrian crossing time. Despite technical problems, the results were considered sufficiently positive to continue development work.

Reading undertook further work on Puffins, into user behavior and pedestrian detection. "The stressful conflict between the pedestrians and vehicles during the flashing amber period is eliminated and replaced by stages of clearly defined priority." (Reading, Dickinson, and Barker, 1995). However, the Puffins and consequently some aspects of the experiments, suffered from inadequate and unreliable equipment. For example, 7.6 percent of valid pedestrian requests were undetected and a number of false cancellations occurred. The potential for using computer vision-based pedestrian detection systems is explored (Reading, Wan, and Dickinson, 1995) as these would (potentially) allow detection of not only the presence but also the volume of pedestrians. However, it was concluded that the vision-based systems available at the time were too restrictive for the Puffin task and that long-term development would be necessary.

Crabtree (1997) describes tests that combine the TRL signal control operating system MOVA (Micro-processor Optimised Vehicle Actuation) (DOT, 1997) and VSPD (Volume Sensitive Pedestrian Detection) with the Puffin (and Pelican) crossing. Two forms of VSPD were used: both used computer processed video images. "The most significant outputs from the project were the VSPD which can sense the numbers of pedestrians waiting to cross and a version of MOVA that can take input from the detector and include it in its optimization process." In terms of delays, MOVA with or without VSPD performed better for both vehicles and pedestrians than VA. The VSPD MOVA with Puffin was more responsive to pedestrian demand but gave greater vehicle delays compared to the MOVA Puffin. At the Puffins, some safety behavior changes were noted. More pedestrians were looking at the traffic rather than straight ahead (where the green man would be located on the Pelican). There were fewer serious crossing infringements, i.e., fewer pedestrians crossed during the green to traffic, "probably because of the reduced delay for all Puffin crossings." However, there were more slight infringements, i.e., pedestrians crossing within the red to traffic but outside the green man period.

Further tests have been carried out, and there are now over 60 Puffin test sites. Despite problems with the reliability of the equipment, the DETR is sufficiently convinced that the Puffin should replace the Pelican. The combination of Puffin and MOVA and possibly VSPD makes the Puffin responsive to local conditions and gives the potential to adjust priorities to suit local policies. Regulations enabling local authorities to install Puffin crossings without Government approval came into force in December 1997. Performance specifications have been issued to industry, and it is believed that suitable on-crossing detection systems and pedestrian demand units (incorporating the push-button and near-side signal) will be produced in the near future. (The performance specification does not stipulate the type of technology.) A Pelican costs approximately £10-15,000 ($16,000 - $25,000) at 1997 prices and a Puffin costs approximately £2,000-3,000 ($3000 - $5000) more than that.

6.3   Toucan crossings

As cyclists are not permitted to cycle across Zebra or Pelican crossings, the Toucan crossing is designed for shared use by pedestrians and cyclists. During the 1980s, the DOT developed a parallel signal-controlled crossing for pedestrians and cyclists. However, this proved to have limited applications as it was expensive, required considerable curb-side space and was generally considered to be an overkill. Trevelyan and Ginger (1989) found that where cyclists cycled over Zebra or Pelican crossings, there were no safety or practical problems for pedestrians. As a result,
a shared crossing — the Toucan — was developed (Morgan, 1993). This has a red man/green man and a green bicycle aspect on a single far-side pole. It is push-button operated often with additional vehicle actuation for pedal cycles. There are now over 200 Toucan crossings in the United Kingdom although they still require special authorization from the DETR. More recent installations include infra-red on-crossing detection and nearside aspects, like the Puffin. Further research has been undertaken by TRL for the DETR into various technical specifications and user behavior (Taylor and Halliday, 1997). As with the Puffins, there have been problems with the reliability of the equipment but user response (from pedestrians and cyclists) has been favorable. It looks likely that future versions of the Toucan will be very similar to the Puffin but with cycle aspects too.

6.4   Further developments in signal-controlled pedestrian crossings

Hunt and Lyons (1997) and others have been critical of the operating strategies for signal-controlled pedestrian crossings (including those at traffic signal junctions) because of the delays to pedestrians and the consequent risks that arise from crossing outside the green man period. They suggests various improvements to reduce pedestrian delays, including:

> Radical approach: At sites where pedestrian activity is high and vehicle occupants are to be given low priority, the signals should default to red-to-vehicles rather than (at present) red-to-pedestrians

> Balanced approach: Allowing the pedestrian precedence period to start when there are five vehicles or less in the detection zones 25-150 m (82 - 492 ft) upstream. Model simulations of this approach suggest lower cycle times, reduced pedestrian delay and crossing during red man, and small increases in vehicle delay.

6.5   Signal control at junctions

It is now usual to include pedestrian crossing facilities within busy signal-controlled junctions. Where traffic signals have been introduced to control traffic at roundabouts (usually caused by capacity problems), these have also proved beneficial for pedestrians.

7.   Footways

7.1   Footway standards

Footways — sidewalks in the United States — are defined as a path for pedestrians <u>adjacent</u> to the carriageway. They are commonly referred to by the public in the United Kingdom as pavements. They are provided on nearly all roads in urban areas. However, the standard of provision varies considerably with the age of the town, the function of the street, the townscape, and the local highway practice. Roads running through rural villages often have no footways for historic reasons.

There is no dedicated manual of UK footway design. Footway standards tend to be included in highway design standards and guidance. Transport in the Urban Environment (IHT, 1997, Chapter 22) recommends that all pedestrian footways should have a minimum width of 1.8 m   (6 ft) but should be wider wherever possible. Dropped crossings (i.e., dropped curbs) should be provided where pedestrians with push-chairs (buggies) or wheelchairs are likely to need to cross. There should be no vertical face on the upstand. The gradients of ramps should be not greater than 8 percent (1:12) but a gradient of 5 percent (1:20) is preferred. Additional references on technical standards for footway design are also given.

May and Hopkinson (1991) undertook systematic studies to assess pedestrians' perception of their environment. Leake et al (1991) assessed ergonomic issues affecting disabled pedestrians.

7.2   Pedestrian route networks

York is a UK city with a particularly high level of walking (24% of journeys to work), and it has sought to consolidate and increase use of this mode. Part of its strategy has been the identification of a 120 km (76 mi) Pedestrian Route Network where higher standards for pedestrian facilities will apply. These include:

- A minimum footway width of 1.8 m (6 ft) increasing to 4.5 m (15 ft) at shops.
- Higher maintenance and lighting standards.
- Traffic calming with at grade crossings.

The cost of implementing these measures is estimated at £1million ($1.6 million) spread over 10 years (White, 1994).

7.3   Maintenance

The issue of footway maintenance has received considerable attention recently, because of budget cuts, concerns that footway standards are declining, and increases in injury claims. Parking on the footway, overruns by vehicles, and work taking place in the footway are all increasing problems for various reasons, including pedestrian safety and amenity. Reinstatement of highway works on utilities (water, electricity, etc.) must now be carried out in accordance with a national specification (DOT, 1992b).

National footway maintenance standards are regularly monitored (DOT, 1996d). A major study (Burtwell (Ed), 1995) reviewed management methods, design and construction, the causes of footway problems, repair and monitoring techniques, whole life costing, the needs of the mobility handicapped, and current research. A follow-up study into pedestrian attitudes has also been undertaken.

7.4   Tactile paving surfaces

Tactile paving is now widely used to help alert and guide visually-impaired pedestrians on the footway. The DETR (1997d) has recently completed a comprehensive guidance note on the use of tactile paving surfaces. It covers surfaces to guide visually-impaired people at crossings, (blister paving), hazard warning surfaces, (corduroy paving), tactile lines to segregate cyclists and pedestrians on shared-use routes, and other forms of warning and information surfaces. It confirms (section 1.1) earlier guidance that, at dropped curbs, there should be no vertical upstand as even a minimal vertical upstand can be a hazard to wheelchair users. Tactile paving should be used to assist visually-impaired pedestrians, as a curb upstand of less than 25 mm is insufficient for them or their guide dogs to detect. Revised guidelines on providing for pedestrians with mobility impairment are provided by the IHT (1991). (See figure 10.)

7.5   Pedestrians and cyclists

Although there are few reported accidents between cyclists and pedestrians, there is growing concern amongst pedestrians, particularly visually-impaired people, about cyclists riding (illegally) on footways and shared-use schemes where part of the footway is converted to a cycle track. Pedestrian and cyclist organizations agreed that cycle routes should normally be provided by redistributing space from cars, not pedestrians (CTC & PA, 1995). There are calls for a moratorium on further shared-use schemes (Bendixson, 1997), and the DETR is reviewing its previous guidance (DOT, 1986). Tests have been undertaken to try to establish the best form of tactile white line delineator to separate the cycle track from the footway. It was concluded (Savill et al, 1997) that the currently authorized profile line (to Dia 1049.1) (DOT, 1990) was the design most easily detected by visually-impaired pedestrians using canes, without causing a hazard to other users. This has a height of 12 to 20 mm with sloping shoulders; the 20 mm height is recommended.

7.6   Other footway issues

Cullen (1997) provides an comprehensive overview of shared-use issues, not just cyclists, from the pedestrian's perspective. A significant and growing problem is parking on the footway (Pickett, 1995). Binns (1991) reviews the experience of the London footway parking ban.

7.7   Pedestrian guard rail

Pedestrian guard rail is sometimes used to prevent pedestrians from crossing at locations deemed particularly hazardous, such as at junctions or roads with high-speed traffic. It is also used to channel pedestrians towards designated crossing points. Safety auditors will often insist on its

Figure 10. Tactile warning in a curb ramp.

installation in new schemes. However, guard rails are generally unpopular with pedestrians as it restricts their freedom to cross and narrows the usable width of the footway. (See figure 11.) It is also disliked because most designs are unattractive and obtrusive in the street scene. Rather oddly (considering how widely it is used), there is very little research into its effectiveness. Stewart (1988) found that the traditional guard rail reduced adult pedestrian casualties but increased child pedestrian casualties because children were masked from drivers by the railings. He developed Visirail which permits drivers to see through the railings.

8.   Traffic Calming, Speed Reduction, and Pedestrians

8.1   Recent UK traffic calming

Traffic calming has been the focus of much research, development, and implementation in the past decade as the United Kingdom has sought to catch up in this field with other European countries such as Holland, Germany, and Denmark. The Traffic Calming Act 1992, The Highways (Traffic Calming) Regulations 1993, and The Highways (Road Humps) Regulations 1996 have clarified and increased the powers of UK highway authorities to introduce traffic calming measures and to be more innovative. More recently, 32 km/h (20 mi/h) speed limit zones have been introduced where the lower speed limit is enforced by physical measures. The DETR allocates some £60 million ($98 million) per annual to local highway authorities in England for Local Safety Schemes and a substantial proportion of this now goes for traffic calming schemes (Tootill and Mackie, 1995).

Traffic calming refers to a wide range of physical measures, including vertical and horizontal deflections, changed surfaces, planting, etc. It may include specific facilities for pedestrians, such as pedestrian refuges, curb build-outs, entry treatments, or Zebra crossings; and these features are reviewed in section 5 above. Although traffic calming is not a pedestrian facility, it can be parti- cularly beneficial to pedestrians as they are most vulnerable to excessive speed even on 48 km/h (30 mi/h) roads which was until recently the lowest permitted speed limit for a public road.

The standard height for road humps is now set at 75 mm which has been found to be just as effective at reducing vehicle speeds as the 100-mm humps (Webster and Layfield, 1996). Flat-top humps provide easier crossings for pedestrians and can be combined with Zebra crossings, although they are generally more uncomfortable for bus passengers than round-top humps.

Side road entry treatments — usually flat-top hump in combination with curb build-outs and other physical measures — can be provided to help pedestrians cross side roads. This creates greater route continuity for pedestrians and reinforces the Highway Code rule that drivers turning into side roads should give way to pedestrians who are crossing.

Although traffic calming is considered to have substantial casualty reduction effects, the United Kingdom's approach has been criticized for over-reliance on physical measures, particularly road humps, and a lack of attention to aesthetics, changing driver attitudes, or integration with wider transport and environmental policy objectives. Some measures have had adverse consequences for emergency services, bus users, and cyclists.

Figure 11. Examples of pedestrian guardrails.

8.2   32 km/h (20 mi/h) zones

Thirty-two km/h (20 mi/h) zones appear to be highly successful at reducing vehicle speeds and casualties. Webster and Mackie (1996) found that the average all accident frequency fell by 60 percent and that child pedestrian accidents were reduced by 70 percent. There was a 6.2 percent reduction in accidents for each 1.61 km/h (1 mi/h) reduction in vehicle speed. At the time of reporting, there were over 200 32 km/h (20 mi/h) zones in the United Kingdom of which 82 had been granted permanent status. (See figure 12.) While this seems a very encouraging initiative, it must be remembered that the 32 km/h (20 mi/h) zones are relatively small areas, excluding linear sites, the average size of a 32 km/h (20 mi/h) zone is 0.28 km and the cost of applying these techniques to the majority of residential roads is prohibitive. PACTS (1996) and others have suggested trials of 32 km/h (20 mi/h) limits without fully-enforcing physical measures, equivalent to the 30 km/h (19 mi/h) limits adopted throughout the city of Graz, Austria, on all roads except a few strategic routes where 50 km/h (31 mi/h) limits apply.

8.3   Other speed reducing initiatives

Some UK local highway authorities are developing speed management plans. York (Pheby, 1997) found strong public support for a plan which categorizes all roads into traffic, mixed priority, or residential, with appropriate speed management and enforcement policies. The city of Gloucester has been selected for the 5-year Safe City project which involves a combination of physical and other safety measures, including speed management, on the basis of the Urban Safety Management guidelines (IHT, 1990).

A range of ambitious measures, including lower speed limits and speed governors in vehicles are proposed by PACTS (1996), the European Transport Safety Council (1995), and Plowden and Hillman (1996). However, lowering speed limits in urban areas does not feature as an option in the DETR's recent review of road safety strategy (DETR, 1997b).

9.   School Zone Safety

9.1   Safe routes to schools

There has recently been increased attention by DETR, local authorities, and non-governmental organizations given to improving the safety of routes to schools for child pedestrians and cyclists (Wood, 1995). This typically involves a combination of traffic calming techniques, provision of crossings, and shared-use pedestrian and cyclist paths. As reported accidents on the school journey are relatively rare (Hillman et al, 1991), this tends to address fears about traffic danger along the route and the difficulties of crossing busy roads, rather than specific accident problems. In addition, 32 km/h (20 mi/h) zones, while not providing specific routes to schools, have proved effective at reducing child casualties and increasing the confidence of parents and children to walk (Webster and Mackie, 1996). Clarke (1997) has positively evaluated the pilot Safe Routes to Schools projects initiated by the charity Sustrans.

Figure 12. Thirty-two km/h (20mi/h) zone on United Kingdom street.

9.2 Other measures

Variable message signs have been tested in the vicinity of schools to warn drivers of excessive speed. Although these have shown some speed reducing effects, they are expensive and less effective compared to physical traffic calming measures and therefore generally considered unsuitable.

10. Education

10.1 Child pedestrians

The DETR's child pedestrian safety strategy (DOT, 1996b) continues the Government's commitment to support child road safety education while placing more emphasis on driver responsibility and speed reduction through physical measures and advertising campaigns.
Research commissioned by the DETR (Thompson et al, 1996) concludes that the commonly-held model of child psychological development, which assumes that children are incapable of understanding traffic until the age of about 9, is erroneous and that given proper methods, children can be taught effective road safety skills at a young age (from around 5 years old) and do not need to be separated from traffic until their "childhood has matured out of them" at the age of about 9.

10.2 Drivers

The effects of vehicle speed on the severity of pedestrian casualties has been a recurring theme (DOT, 1992a). The DETR has continued and increased the "shock index"level of its Kill your Speed TV adverts, employing home video of actual child victims. These stress appropriate speeds rather than simply staying within speed limits. The message also indicates that drivers have a special responsibility for the safety of children.

Some local authorities, e.g., Suffolk, have introduced "speed pledge" campaigns whereby drivers are invited to sign the pledge to drive at appropriate speeds and to display car stickers to that effect. In some cases this has been in cooperation with the police who have invited speeding motorists to sign the pledge.

Unfortunately, there seems to be growing evidence (Silcock et al 1993) that drivers are concentrating on other motor vehicles — which represent a potential threat to them — and not on pedestrians. This may be a reflection of the increased safety of car occupants relative to pedestrians, and increased levels of motorized traffic. The authors conclude that the worst drivers are unlikely to respond to education alone.

11. Enforcement

Probably the most significant factor in this area is the use of GATSO automatic speed cameras which has increased substantially in the past 5 years because of technical, legal, and administrative progress, linked to the issuing of fixed penalty notices. They have been shown to be effective at reducing vehicle speeds and accidents at specific locations but, so far, not more widely. It is likely that their use will increase substantially, particularly if funding problems can be resolved. Mobile speed cameras are also now employed by the police (Pheby, 1997).

12.   Conclusions

The past 5 years has seen increased attention given to road safety issues in the United Kingdom. Developments of particular relevance to pedestrians include greater emphasis on reducing vehicle speeds in urban areas through physical, legal, and publicity measures; also development of the Puffin crossing and new operating strategies such as MOVA. However, while specific facilities can affect safety at individual sites, improvements in overall safety for pedestrians require a comprehensive road safety strategy that is fully integrated with land use and transport policy (DETR, 1997e). Amendments to the Construction and Use regulations for motor vehicles, greater emphasis on driver responsibility towards pedestrians, and reductions in traffic levels will also be needed to bring about further accident reductions and a perception that walking is becoming safer.

13.   References

Adams, J (1985) *Risk and Freedom - The Record Of Road Safety Legislation*

Bendixson, T (1997) "Walking to Happiness," *Surveyor*, 2 October, pp 22-23

Billings K and B Walsh (1991) "New Pedestrian Facilities at Signalled Junctions," PTRC Seminar K proceedings, Vol P350, pp 1-12

Binns, R (1991) "The GLC Pavement Parking Ban and What Happened To It" *Walk*, Vol 6, No 15, pp 12-20, Pedestrians Association, London

Burtwell, MH (Ed) (1995) *A Study of Footway Maintenance*, TRL Report 134 Transport Research Laboratory, Crowthorne

Carsten, OMJ, Tight, MR, Southwell, MT with Plows, B (1989) *Urban Accidents: Why do They Happen?* AA Foundation for Road Safety Research, Basingstoke

Christie, N (1995) *The High Risk Child Pedestrian: Socio-economic and Environmental Factors in their Accidents*. TRL Project report 117, Transport Research Laboratory, Crowthorne

Clarke, S (1997) *Sustrans Safe Routes to Schools Project - An Evaluation,* Sustrans, Bristol

Crabtree, MR (1997) "A Study of Four Styles of Pedestrian Crossings," PTRC Proceedings Seminar K Vol, P419, London

CSS (1997) *The Road Safety Record of Pedestrian Crossings in Relation to PV^2*. CSS (Formerly County Surveyors Society), Gloucester

Cullen, P (1997) "To Share or Not to Share?" *Walk* Vol 8, No 3. Pedestrians Association, London

Cyclists' Touring Club and Pedestrians Association (1995) *Joint Statement on Providing for Walking and Cycling as Transport and Travel* CTC, Godalming

Davies, DG, TJ Ryley, SB Taylor, and ME Halliday (1997) *Cyclists at Road Narrowings*, TRL Report 241, Transport Research Laboratory, Crowthorne

Davies, HEH (1992) *The Puffin Crossing: Experience with the First Experimental Sites*, TRL Research Report 364 Transport Research Laboratory, Crowthorne

Davies, H and MA Winnett (1993) *Why do Pedestrian Accidents Happen?* PTRC Proceedings, Vol P365, pp 315-24, London

Davis, R (1993) *Death on the Streets*, Leading Edge Press, North Yorkshire

Department of the Environment and Department of Transport (1992) *Design Bulletin 32. Residential Roads and Footpaths Layout Considerations.* HMSO, London

Department of the Environment and Department of Transport (1994) *Planning Policy Guidance Note 13 (PPG13) Transport* HMSO, London

Department of the Environment, Transport and the Regions (1997a) *Road Accidents Great Britain 1996: The Casualty Report* The Stationery Office, London

Department of the Environment, Transport and the Regions (1997b) *Towards Safer Roads* Department of the Environment, Transport and the Regions, London

Department of the Environment, Transport and the Regions (1997c) *Road Safety Strategy. Current Problems and Future Solutions,* Department of the Environment, Transport and the Regions, London

Department of the Environment, Transport and the Regions (1997d) *Guidance on the Use of Tactile Paving.* (Notified Draft, September 1997) Department of the Environment, Transport and the Regions, London.

Department of the Environment, Transport and the Regions (1997e) *Developing an Integrated Transport Policy — Factual Background,* Department of the Environment, Transport and the Regions, London

Department of Transport (1986) Local Transport Note 2/86 *Shared Use by Cyclists and Pedestrians,* HMSO, London

Department of Transport (1987a) *Road Safety: The Next Steps (and Detailed Review),* Department of Transport, London

Department of Transport (1987b) Technical Advice Note TA 52/87 *Pelican and Zebra Crossings,* Department of Transport, London

Department of Transport (1990) Traffic Advisory Leaflet 4/90, *Tactile Markings for Segregated Shared Use by Cyclists and Pedestrians,* Department of Transport, London

Department of Transport (1991) *The History of Traffic Signs,* Department of Transport, London

Department of Transport (1992a) *Killing Speed and Saving Lives. The Government's Strategy for Tackling the Problem of Excess Speed on Our Roads,* Department of Transport, London

Department of Transport (1992b) *New Roads and Street Works Act 1991. Specification for the Reinstatement of Openings in Highways. A Code of Practice.* HMSO, London

Department of Transport (1995a) *Road Safety Report 1995,* Department of Transport, London

Department of Transport (1995b) *Design Manual for Roads and Bridges*, HMSO, London

Department of Transport (1995c) Local Transport Note 1/95 *The Assessment of Pedestrian Crossings*, HMSO, London

Department of Transport (1995d) Local Transport Note 2/95 *The Design of Pedestrian Crossings*, HMSO, London

Department of Transport (1996a) *Developing a Strategy for Walking*, Department of Transport, London

Department of Transport (1996b) *Child Pedestrian Safety in the UK. A Strategy for Reducing Child Pedestrian Casualties*, Department of Transport, London

Department of Transport (1996c) *Transport Statistics Great Britain: 1996 Edition.* HMSO, London

Department of Transport (1996d) *National Road Maintenance Condition Survey.* Statistics Bulletin 96 (30) Department of Transport, London

Department of Transport (1997) *Traffic Advisory Leaflet 3/97. The MOVA Signal Control System*, Department of Transport, London

Elvik, R (1997) "Evaluations of Road Accident Blackspot Treatment: A Case of the Iron Law of Evaluation Studies?" *Accident Analysis and Prevention*, Vol 29, No 2, pp 191-199

European Transport Safety Council (1995) *Reducing Traffic Injuries from Excess and Inappropriate Speed*, European Transport Safety Council, Brussels

Hillman, M, Adams, J and Whitelegg, J (1991) *One False Move... A Study of Children's Independent Mobility.* PSI

Hopkin, JM, PA Murray, M Pitcher and CSB Galasko (1993) *Police and Hospital Recording of Non-fatal Road Accident Casualties: A Study in Greater Manchester*, TRL Research Report 379, Transport Research Laboratory, Crowthorne

House of Commons Transport Committee (1996a) Third Report: *Risk Reduction for Vulnerable Road Users*, HMSO, London

House of Commons Transport Committee (1996b) Session 1996-97. First Special Report: *Government Observations on the Third Report of the Committee (Risk Reduction for Vulnerable Road Users)* The Stationery Office, London

Hunt, JG (1997) "Pedestrian Crossings — Changing the Balance of Priorities," PTRC pp 325-337

Hunt, JG and GD Lyons (1997) "Enhanced Operating Strategies to Improve Pedestrian Amenity and Safety at Midblock-Signalled Pedestrian Crossings," PTRC Vol P 419, pp 183-196

Institution of Highways and Transportation (1990) *Guidelines on Urban Safety Management*, IHT, London

Institution of Highways and Transportation (1991) *Reducing Mobility Handicaps: Towards a Barrier-Free Environment. IHT Revised Guidelines* IHT, London

Institution of Highways and Transportation (1996) *Guidelines for the Safety Audit of Highway*, IHT, London

Institution of Highways and Transportation (1997) *Transport in the Urban Environment*, IHT, London

Kindred Associations (1995) *Report on Highways Liability Claims — The Issues, 1995* Merlin Communications (UK) Ltd, Cirencester

Lambert, S (1997) "Progress — But a Great Deal More Needed" *Care on the Road*, April 1997, p 16, Royal Society for the Prevention of Accidents, Birmingham

Lawson, SD (1990) *Accidents to Young Pedestrians: Distributions, Circumstances, Consequences and Scope for Countermeasures*, AA Foundation for Road Safety Research, Basingstoke

Leake, GR, AD May, and T Parry (1991) An Ergonomic Study of Pedestrian Areas for Disabled People" TRRL Contractor Report 184, Transport Research Laboratory, Crowthorne

Local Transport Today (1997) "Government to Set Three New Road Safety Targets for 2010." Issue 223, 23 October 1997, p 5

London Planning Advisory Committee (1996) *Putting London Back on Its Feet. A Strategy for Walking in London.* London Planning Advisory Committee, London

Lynam, D and Harland, G (1992) "Child Pedestrian Safety in the UK," TRL published article, Transport Research Laboratory, Crowthorne

Mackay, M (1994) "Engineering in Accidents: Vehicle Design and Injuries" *Injury*, 1994/10 25(9) pp 615-21

May, AD and Hopkinson, PG *Perceptions of the Pedestrian Environment*, (1991) TRRL Contractor Report 148 Transport and Road Research Laboratory, Crowthorne

Morgan, J (1993) Toucan Crossings for Cyclists and Pedestrians, TRL Project Report 47, Transport Research Laboratory, Crowthorne

O'Reilly, D (1994) "Child Pedestrian Safety in Great Britain" In Department of Transport (1994) *Road Accidents Great Britain: 1993*, pp 33-40, HMSO, London

Parliamentary Advisory Council for Transport Safety (PACTS) (1995) *Targets 2001. Where Do We Go From Here?* Parliamentary Advisory Council for Transport Safety, London

Parliamentary Advisory Council for Transport Safety (PACTS) (1996) *Taking Action on Speeding*, Parliamentary Advisory Council for Transport Safety, London

Pheby, T (1997) *Killing Speed — Saving Lives. The Outcome of Consulting on York's Speed Management Plan.* PTRC Proceedings Vol P419, pp 317-329

Pickett, MW (1995) "Parking Control: From Principle to Practice. A Good Start but Problems with Footway Parking." Proceedings of TRL Parking Seminar. TRL published article PA 3056/65 pp46-69. Transport Research Laboratory, Crowthorne

Plowden, S and M Hillman (1996) *Speed Control and Transport Policy*, PSI, London

Reading, IAD, KW Dickinson and DJ Barker (1995) "The Puffin Pedestrian Crossing: Pedestrian Behavioural Study" *Traffic Engineering + Control,* Sept, pp 472-478

Reading, IAD, Wan CL, Dickinson, KW (1995) "Developments in Pedestrian Detection" *Traffic Engineering + Control* 1995/10, pp 538-542

Royal Commission on Environmental Pollution (1994) *Transport and the Environment*, CM 2674 HMSO, London

Savill, T, C Gallon and G McHardy (1997) *Delineation for Cyclists and Visually Impaired Pedestrians on Segregated Shared Routes*, TRL Report 287, Transport Research Laboratory, Crowthorne

Silcock, D (1993) "Risk on the roads. 3: Links Between Attitudes and Perceptions of Rrisk and Applications of the Results," *Traffic Engineering + Control* 1993/07/08, pp 372-375

Stewart, D (1988) "Pedestrian Guardrails and Accidents" *Traffic Engineering + Control*, September 1988

Taylor SB and ME Halliday (1997) *Pedestrians and Cyclists Attitudes to Toucan Crossings*, TRL Report 277, Transport Research Laboratory, Crowthorne

Thompson, J et al (1996) *Child Development and the Aims of Roads Safety Education*, HMSO, London

Thompson, SJ and S Heydon (1991) "Improving Pedestrian Conspicuity by Use of a Promontory," *Traffic Engineering + Control* 1991/08, pp 472-478

Thompson, SJ, S Heydon and C Charnley (1990) "Pedestrian Refuge Schemes in Nottingham," *Traffic Engineering + Control* 1990/03, pp 118-23

Tootill, W and Mackie, A (1995) *Transport Supplementary Grant for Safety Schemes — Local Authorities' Schemes from 1992/93* TRL Report 127, Transport Research Laboratory, Crowthorne

Trevelyan, P and Ginger, M (1989) *Cyclists Use of Pedestrian and Cycle/Pedestrian Crossings*, TRL Report CR 173, Transport Research Laboratory, Crowthorne

Ward, H, J Cave, A Morrison, R Allsop, and A Evans (1994) *Pedestrian Activity and Accident Risk*, AA Foundation for Road Safety Research, Basingstoke

Webster, DC and RE Layfield (1996) *Traffic Calming — Road Hump Schemes Using 75 mm High Humps*, TRL Report 186, Transport Research Laboratory, Crowthorne

Webster, DC and AM Mackie (1996) *Review of Traffic Calming Schemes in 20 mph Zones*, TRL Report 215, Transport Research Laboratory, Crowthorne

White, J (1994) "A Walk on the (not so) Wild Side — Promoting the Pedestrian in York," PTRC Proceedings Vol P381, pp 1-13, London

Wood, C (1995) "Taking New Routes to School Safety" *Urban Street Environment*, 1995/08/09 pp 11-15, Landor Publishing, London

Zegeer, CV, M Cynecki, J Fegan, B Gilleran, P Lagerway, H C Tan, and R Works, (1994) "Summary Report on FHWA Study Tour for Pedestrian and Bicyclist Safety in England, Germany and The Netherlands" Report No. FHWA-PL-95-006, Federal Highways Administration, Washington DC, USA

14. Glossary of Terms and Abbreviations

DETR   Department of the Environment, Transport and the Regions (formerly Department of Transport). UK Government department responsible for transport in England.

DOE   Department of the Environment (now part of Department of the Environment, Transport and the Regions). UK Government department responsible for land use planning, local government, and other matters in England.

DOT   Department of Transport (now part of Department of the Environment, Transport and the Regions). UK Government department responsible for transport in England.

DRIVE   A transport technology research program funded by the European Community.

IHT   Institution of Highways and Transportation. UK professional association for traffic engineers.

LTN   Local Transport Note. Technical advice note issued by DETR.

MOVA   Micro-processor Optimized Vehicle Actuation. A sophisticated traffic signal control system for isolated junctions or crossings, developed by TRL on behalf of the DOT, that adjusts signal timings in response to demands.

Pelican   PEdestrian LIght CONtrolled crossing. A mid-block signal-controlled crossing for pedestrians.

Puffin   Pedestrian User Friendly INtelligent crossing. A mid-block signal-controlled crossing for pedestrians, that adjusts waiting and crossing times in response to demands. The replacement for the Pelican.

STATS 19   Form used by police in UK to record road traffic accidents.

Toucan   Two Can Cross. A signal-controlled crossing for shared use by cyclists and pedestrians.

TRL   Transport Research Laboratory. Principal UK transport research centre, previously part of DOT but now independent.

TRO (Traffic Regulation Order)   Legal procedure whereby local highway authorities introduce specific traffic controls, such as "No Entry" or "One-Way" restrictions.

UTC   Urban Traffic Control.

VA   Vehicle Actuation. In systems with VA, traffic signals detect and respond to approaching traffic. (In fixed-time systems, traffic signals are not responsive to traffic flow.)

VSPD   Volume Sensitive Pedestrian Detection. A system that can detect and quantify the presence of pedestrians (at signal-controlled pedestrian crossings).

Zebra   Pedestrian crossing without signal control, indicated by black and white bands on the surface of the carriageway. Motor vehicles must give way to pedestrians who have established precedence on the crossing.

Notes

1.	The United Kingdom (UK) comprises England, Scotland, Wales and Northern Ireland. Great Britain (GB) comprises England, Scotland, and Wales. While many aspects of pedestrian facilities are common throughout the four countries, there are legal and administrative differences. This report refers to the situation in England.

2.	Following the UK General Election in May 1997, the Department of Transport (DOT) was merged with other Government departments to form the Department of the Environment, Transport and the Regions (DETR).

Acknowledgments

David Davies Associates is grateful to all those who assisted with this report, particularly to David Williams, Adrian Waddams, Eric Wyatt and Suku Phull (Department of the Environment, Transport and the Regions), and to Paul Cullen and Rosamund Weatherall (Pedestrians Association). However, they bear no responsibility for the contents of this report.

www.ingramcontent.com/pod-product-compliance
Lightning Source LLC
Chambersburg PA
CBHW081356170526
45166CB00010B/3106

APPENDICES

A. Fontana Hotel Floor Plans Showing Extent of Structural Collapse

B. Fontana Hotel Fire Victims

C. Enlarged Floor Plans Showing Room of Residence and Final Position of Victims

D. Florida Statute on Sprinklers in Hotels

E. Photographs

APPENDIX A

Fontana Hotel Floor Plans
Showing Extent of Structural Collapse

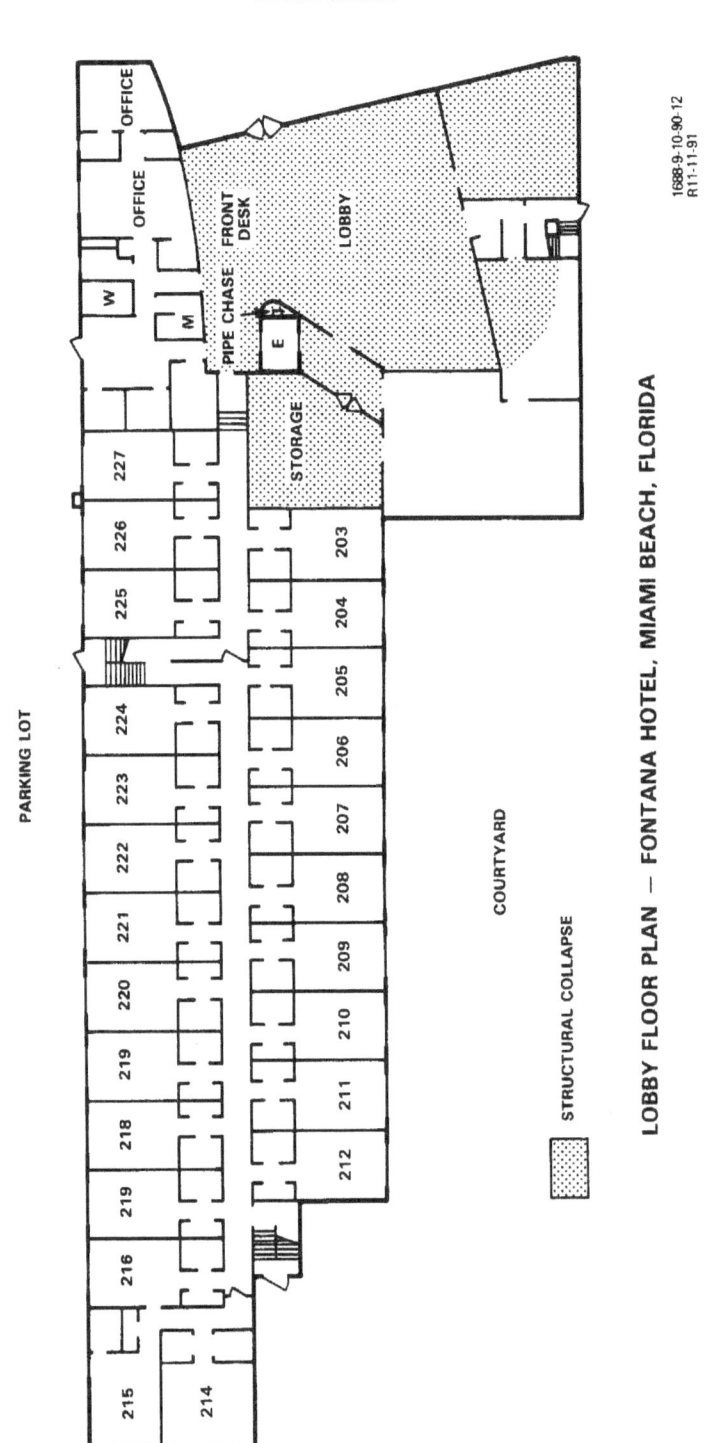

LOBBY FLOOR PLAN — FONTANA HOTEL, MIAMI BEACH, FLORIDA

8

Appendix A (Continued)

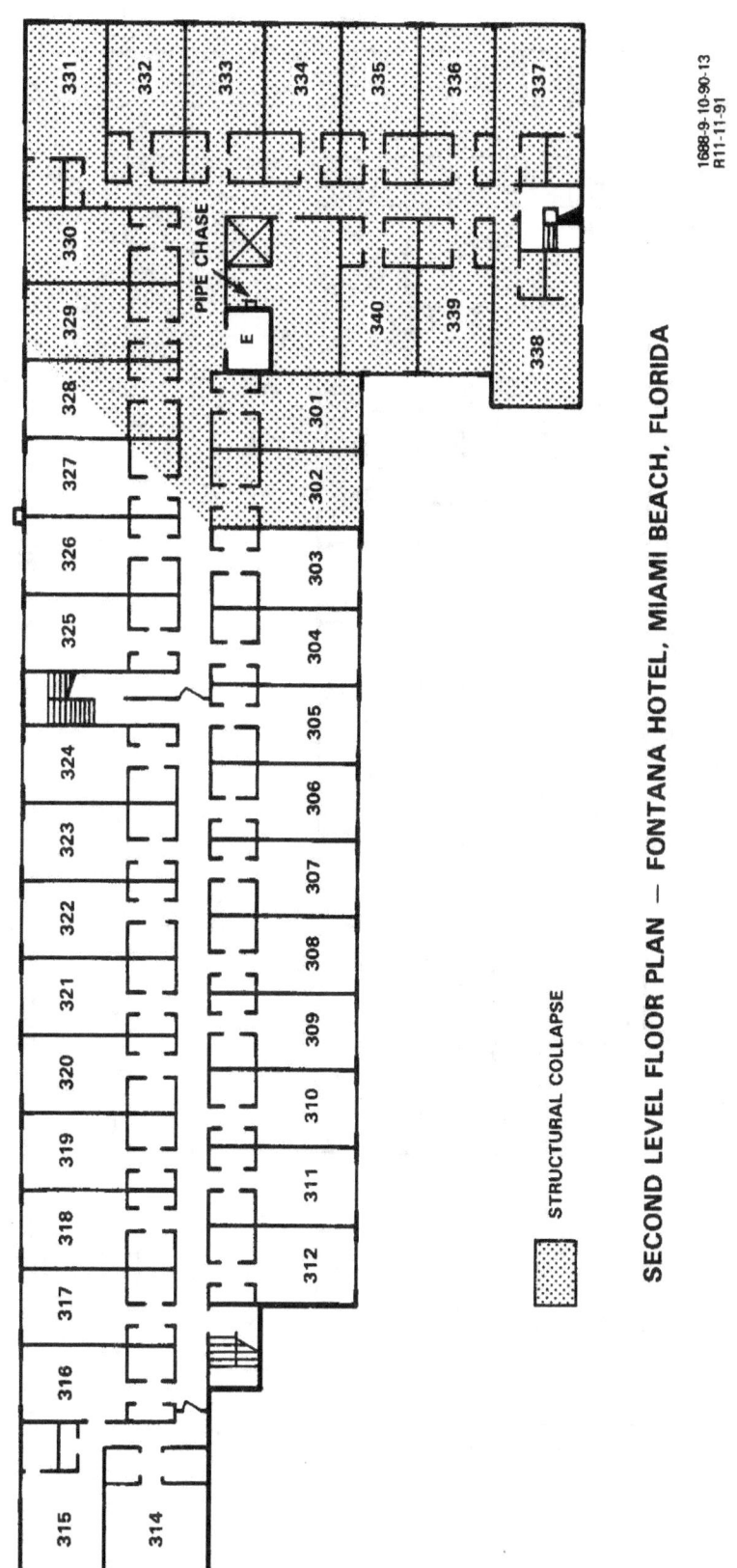

STRUCTURAL COLLAPSE

SECOND LEVEL FLOOR PLAN — FONTANA HOTEL, MIAMI BEACH, FLORIDA

1688-9-10-90-13
R11-11-91

Appendix A (Continued)

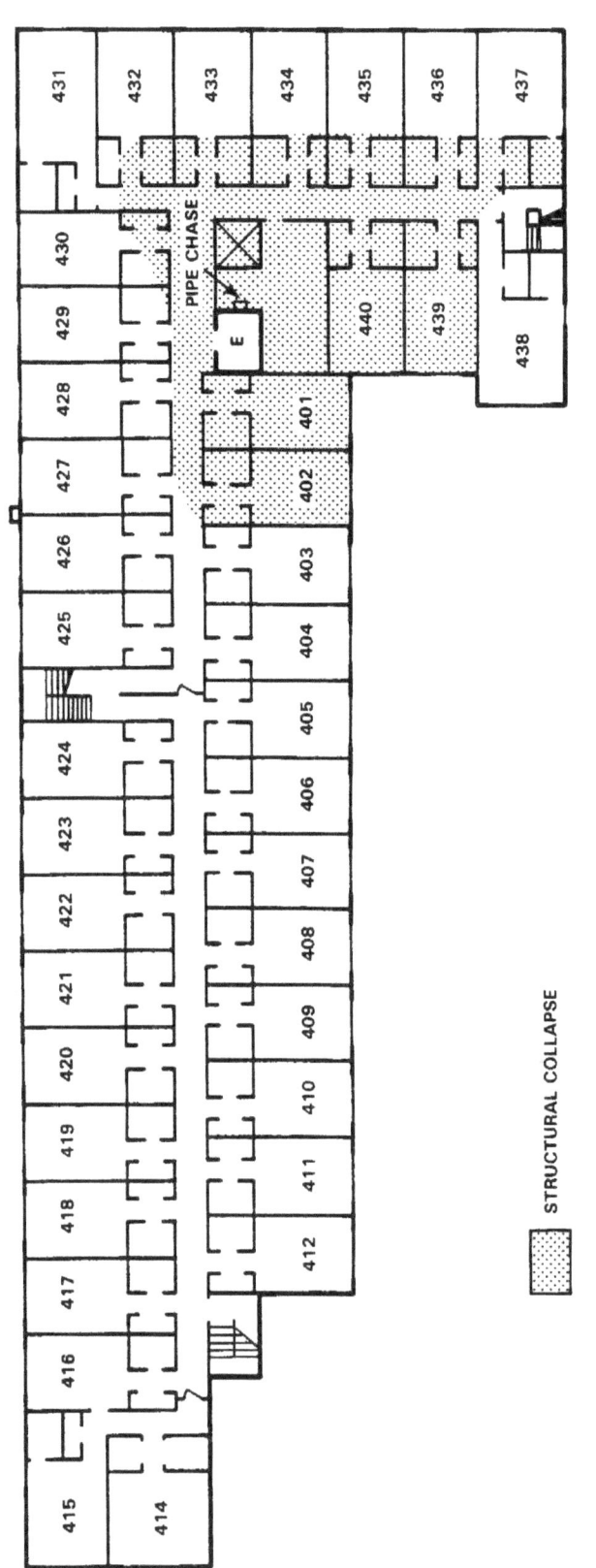

STRUCTURAL COLLAPSE

THIRD LEVEL FLOOR PLAN — FONTANA HOTEL, MIAMI BEACH, FLORIDA

APPENDIX B

Fontana Hotel Fire Victims

	Name	Age	Room Victim Lived In
1.	Sally Mur	75 years	340
2.	Doris Fein aka Dorothy Trapanese	81	439
3.	Selma Rauthaus	85	339
4.	Sara Pollack	87	432
5.	Martha Springer	88	205
6.	Nettie Feingold	89	332
7.	Jack Schwarz	92	330
8.	Dr. Joseph Wels	92	227
9.	Minnie Weinstein	93	433

APPENDIX C

Enlarged Floor Plans Showing Room of Residence and Final Position of Victims

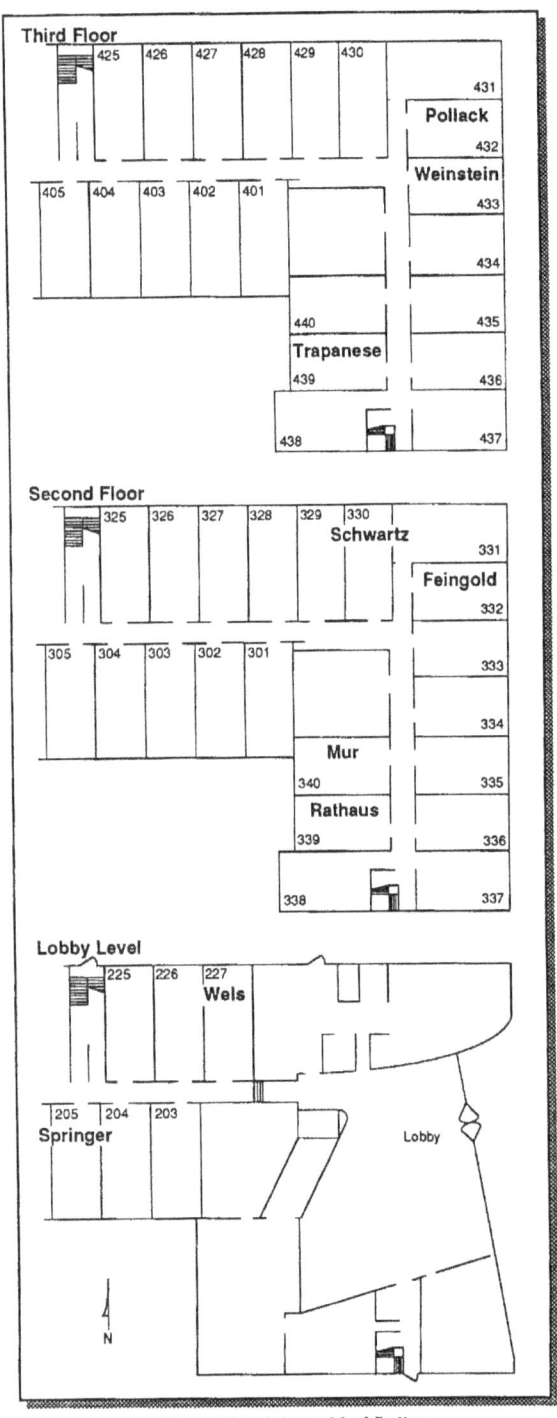

Room Registered to Victim   Location of Bodies

772-5-8-91-1

12

APPENDIX D

Florida Statute on Sprinklers in Hotels

Restaurants shall impose administrative sanctions pursuant to s. 509.261.

History.—s. 1, ch. 90-242.

509.213 Emergency first aid to choking victims.— [The expiration of this section pursuant to s. 5, ch. 83-241, was nullified by s. 51, ch. 90-339. Repealed effective October 1, 2000, by s. 52, ch. 90-339, and scheduled for review pursuant to s. 11.61.]

¹509.214 Notification of automatic gratuity charge. Every public food service establishment which includes an automatic gratuity or service charge in the price of the meal shall include on the food menu and on the face of the bill provided to the customer notice that an automatic gratuity is included.

History.—s. 1, ch. 86-24; s. 1, ch. 88-16; ss. 21, 52, ch. 90-339
¹Note.—Repealed effective October 1, 2000, by s. 52, ch. 90-339, and scheduled for review pursuant to s. 11.61.

¹509.215 Firesafety.—
(1) Any:
(a) Public lodging establishment, as defined in this chapter, which is of three stories or more and for which the construction contract has been let after September 30, 1983, with interior corridors which do not have direct access from the guest area to exterior means of egress, or
(b) Building over 75 feet in height that has direct access from the guest area to exterior means of egress and for which the construction contract has been let after September 30, 1983,

shall be equipped with an automatic sprinkler system installed in compliance with the provisions prescribed in the National Fire Protection Association publication NFPA No. 13 (1985), "Standards for the Installation of Sprinkler Systems." The sprinkler installation may be omitted in closets which are not over 24 square feet in area and in bathrooms which are not over 55 square feet in area, which closets and bathrooms are located in guest rooms. Each guest room shall be equipped with an approved listed single-station smoke detector meeting the minimum requirements of NFPA-74 (1984) "Standards for the Installation, Maintenance and Use of Household Fire Warning Equipment," powered from the building electrical service, notwithstanding the number of stories in the structure, if the contract for construction is let after September 30, 1983. Single-station smoke detection is not required when guest rooms contain smoke detectors connected to a central alarm system which also alarms locally.

(2) Any public lodging establishment, as defined in this chapter, which is of three stories or more and for which the construction contract was let before October 1, 1983, shall be equipped with:
(a) A system which complies with subsection (1); or
(b) An approved sprinkler system for all interior corridors, public areas, storage rooms, closets, kitchen areas, and laundry rooms, less individual guest rooms, if the following conditions are met:
1. There is a minimum 1-hour separation between each guest room and between each guest room and a corridor.

2. The building is constructed of noncombustible materials.
3. The egress conditions meet the requirements of s. 5-3 of the Life Safety Code, NFPA 101 (1985).
4. The building has a complete automatic fire detection system which meets the requirements of NFPA-72A (1987) and NFPA-72E (1984), including smoke detectors in each guest room individually annunciating to a panel at a supervised location.

(3) The Division of State Fire Marshal may prescribe uniform standards for firesafety equipment for public lodging establishments for which the construction contracts were let before October 1, 1983. An entire building shall be equipped as outlined not later than October 1, 1989, except that the approved sprinkler system may be delayed by the Division of State Fire Marshal until October 1, 1991, on a schedule for complete compliance in accordance with rules to be adopted by the Division of State Fire Marshal, which schedule shall include a provision for a 1-year extension which may be granted not more than three times for any individual requesting an extension. The entire system must be installed and operational by October 1, 1994. The Division of State Fire Marshal shall not grant an extension for the approved sprinkler system unless a written request for the extension and a construction work schedule is submitted. The Division of State Fire Marshal may grant an extension upon demonstration that compliance with this section by the date required would impose an extreme hardship and a disproportionate financial impact. Any establishment that has been granted an extension by the Division of State Fire Marshal shall post, in a conspicuous place on the premises, a public notice stating that the establishment has not yet installed the approved sprinkler system required by law.
(4) The provisions for installation of single-station smoke detectors required in subsection (1) and subparagraph (2)(b)4. shall be waived by the Division of State Fire Marshal for any establishment for which the construction contract was let before October 1, 1983, and which is under three stories in height, if each individual guest room is equipped with a smoke detector approved by the Division of State Fire Marshal and the schedule for compliance is not later than October 1, 1986.
(5) Notwithstanding any other provision of law to the contrary, this section applies only to those public lodging establishments in a building wherein more than 50 percent of the units in the building are advertised or held out to the public as available for transient occupancy.
(6) Special exception to the provisions of this section shall be made for hotel structures that are on the National Register of Historic Places as determined by the United States Department of the Interior or that are of historical significance to this state as determined by the State Historic Preservation Officer, designated pursuant to s. 267.061(5), after consultation with the chairperson of the local historic preservation board or commission, if such board or commission exists. For such structures, provisions shall be made for a system of fire protection and life safety support that would meet the intent of the NFPA standards and be acceptable to, and approved by, a commission composed of the director of

Appendix D (Continued)

the Division of Hotels and Restaurants, the director of the Division of State Fire Marshal, and the State Historic Preservation Officer. The director of the Division of State Fire Marshal shall be designated chairperson of the commission and shall record the minutes of each commission meeting.

(7) The Division of State Fire Marshal shall adopt, in accordance with the provisions of chapter 120, any rules necessary for the implementation and enforcement of this section. The Division of State Fire Marshal shall enforce this section in accordance with the provisions of chapter 633, and any establishment licensed under this chapter in violation of this section may be subject to administrative sanctions by the division pursuant to s. 509.261.

(8) Specialized smoke detectors for the deaf and hearing-impaired shall be available upon request by guests in public lodging establishments at a rate of at least one such smoke detector per 50 dwelling units or portions thereof, not to exceed five such smoke detectors per public lodging facility.

History.—ss. 1, 3, 4, ch 63-194, s 91 ch 85-81, s 7 ch 86-174, s 32 ch ss 1 ch ch 203 ss 22, 51, 52, ch 90-133

¹**Note.**—repealed effective October 1 2000 by s 52, ch 90-339 and scheduled for review pursuant to s. 11.61

¹509.221 Sanitary regulations.—

(1) Each public lodging establishment and each public food service establishment shall be supplied with potable water and shall provide adequate sanitary facilities for the accommodation of its employees and guests. Such facilities may include, but are not limited to, showers, handwash basins, toilets, and bidets. Such sanitary facilities shall be connected to approved plumbing. Such plumbing shall be sized, installed, and maintained in accordance with applicable state and local plumbing codes. Wastewater or sewage shall be properly treated onsite or discharged into an approved sewage collection and treatment system.

(2) Each public lodging establishment and each public food service establishment shall maintain not less than one public bathroom for each sex, properly designated. Each transiently rented public lodging establishment that does not provide private or connecting bathrooms shall maintain one public bathroom on each floor for every 15 guests, or major fraction of that number, rooming on that floor.

(3) Each establishment licensed under this chapter shall be properly lighted, heated, cooled, and ventilated and shall be operated with strict regard to the health, comfort, and safety of the guests. Such proper lighting shall be construed to apply to both daylight and artificial illumination.

(4) Each bedroom in a public lodging establishment shall have an opening to the outside of the building, air shafts, or courts sufficient to provide adequate ventilation. Where ventilation is provided mechanically, the system shall be capable of providing at least two air changes per hour in all areas served. Where ventilation is provided by windows, each room shall have at least one window opening directly to the outside.

(5) Each public lodging establishment renting transiently and each public food service establishment shall provide in the main public bathroom soap and clean towels or other approved hand-drying devices and each public lodging establishment shall furnish each guest with two clean individual towels so that two guests will not be required to use the same towel unless it has first been laundered.

(6) Each public lodging establishment renting transiently shall provide each bed, bunk, cot, or other sleeping place for the use of guests with clean pillowslips and under and top sheets. Sheets and pillowslips shall be laundered before they are used by another guest, a clean set being furnished each succeeding guest. All bedding, including mattresses, quilts, blankets, pillows, sheets, and comforters, shall be thoroughly aired, disinfected, and kept clean. Bedding, including mattresses, quilts, blankets, pillows, sheets, or comforters, may not be used if they are worn out or unfit for further use.

(7) The operator of any establishment licensed under this chapter shall take effective measures to protect the establishment against the entrance and the breeding on the premises of all vermin. Any room in such establishment infested with such vermin shall be fumigated, disinfected, renovated, or other corrective action taken until the vermin are exterminated.

(8) A person, while suffering from any contagious or communicable disease, while a carrier of such disease, or while afflicted with boils or infected wounds or sores, may not be employed by any establishment licensed under this chapter, in any capacity whereby there is a likelihood such disease could be transmitted to other individuals. An operator that has reason to believe that an employee may present a public health risk shall immediately notify the proper health authority.

(9) Subsections (2), (5), and (6) do not apply to any facility or unit classified as a resort condominium as described in s. 509.242(1)(c).

History.—ss. 12, 16, 24, 26, 32, ch 6952, ss. 1, 5, ch 6953, 1915, RGS 2132, 2136, 2144, 2146, 2152, 2156, 5612, ss. 5, 6, 10, ch 9064, 1923, ss. 3, 4, ch 12053, 1927, CGL 3361, 3365, 3373, 3375, 3381-3385, 7830, ss. 14, 18, 26, 28, 34-37, ch 16042, 1933, CGL 1936 Supp. 3361-3365, 3373-3375, 3381, 3382, 3384, 3385, s. 8, ch 57-369, s. 1, ch 59-152, ss. 16, 35, ch 69-106, s. 3, ch 71-157, s. 18, ch 73-325, s. 3, ch 76-168, s. 1, ch 77-174, s. 1, ch 77-457, ss. 17, 39, 42, ch 79-240, ss. 2, 4, ch 81-161, s. 388, ch 81-259, ss. 2, 3, ch 81-318, ss. 3, 4, ch 82-84, ss. 3, 4, ch 83-241, ss. 23, 51, 52, ch 90-339

¹**Note.**—Repealed effective October 1, 2000, by s. 52, ch 90-339, and scheduled for review pursuant to s. 11.61

Note.—Former ss. 511.13-511.17, 511.25-511.27, 511.35-511.37, 511.42.

¹509.232 School carnivals and fairs; exemption from certain food service regulations.—

Any public or non-profit school which operates a carnival, fair, or other celebration, by whatever name known, which is in operation for 3 days or less and which includes the sale and preparation of food and beverages must notify the local county health unit of the proposed event and is exempt from any temporary food service regulations with respect to the requirements for having hot and cold running water; floors which are constructed of tight wood, asphalt, concrete, or other cleanable material; enclosed walls and ceilings with screening; and certain size counter service. A school may not use this notification process to circumvent the license requirements of this chapter.

History.—s 1, ch. 81-147, ss 24, 52, ch 90-339

¹**Note.**—Repealed effective October 1, 2000, by s. 52, ch 90-339, and scheduled for review pursuant to s. 11.61

Appendix D (Continued)

(2) Trouble signals shall be audible and distinctive from alarm signals, and shall comply with NFPA 72-A 2-7.3.

(3) The annunciator system shall have primary power supplied in accordance with NFPA 72-A 2-6.3 and 2-6.7, and secondary power supplied in accordance with NFPA 72-A 2-6.4.

4A-43.011 Extinguishment Requirements - Automatic Fire Sprinklers. Specific Authority 509.215(7), 633.05, 633.051, F.S. Law Implemented 509.215, 633.01, 633.081, F.S. History - New 11-12-85, Formerly 4A-43.10, Repealed_____.

4A-43.012 Standpipe and Hose Systems. Standpipe and hose systems are required for all transient public lodging establishments which are located in buildings exceeding 50 feet in height or in building over six stories high which have a complete automatic sprinkler system. The standpipe and hose systems shall comply with the provisions of NFPA 14, as adopted in Rule Chapter 4A-3.012, Florida Administrative Code. Specific Authority 509.215(7), 633.01, F.S. Law Implemented 509.215, 633.022, F.S. History - New 11-12-85, Formerly 4A-43.11, Amended 8-24-87, _____.

4A-43.013 Places of Assembly. Specific Authority 509.215(7), 633.05, 633.051, F.S. Law Implemented 509.215(7), 633.01, F.S. History - New 11-12-85, Formerly 4A-43.12, Repealed_____.

4A-43.014 Fire Safety Standards for Other Buildings, Equipment or Devices. Specific Authority 509.215(7), 633.05, 633.051, F.S. Law Implemented 509.215(7), 633.01, F.S. History - New 11-12-85, Formerly 4A-43.13, Repealed_____.

4A-43.015 Special Compliance Schedule.

(1) Individuals failing to comply with the October 1, 1989, deadline for sprinklers as stipulated by Section 509.215, and 721.24, Florida Statutes, may be granted an extension of time upon compliance with the following requirements:

(a) The individual shall submit a letter of notification of intent to request an extension to the State Fire Marshal. This letter shall be received in the Office of the State Fire Marshal in Tallahassee no later than 5:00 P.M. on October 1, 1990, and

(b) The individual shall submit an engineering design plan which is in compliance with Chapter 471, Florida Statutes, together with a construction schedule to the State Fire Marshal. The engineering design plan and the construction schedule together with evidence demonstrating that compliance with this section by the date required would impose an extreme hardship and a disproportionate financial impact shall be received in the Office of the State Fire Marshal in Tallahassee no later than 5:00 P.M. on January 1, 1991.

43-4

APPENDIX E

Photographs

Slides and photographs are included with the master report at the USFA. Selected photos are presented on the following pages. They are by the author or were provided by the Miami Beach Fire Department (MBFD) as noted.

Lobby of the Prince Michael Hotel, interior finish and construction is similar to that of the Fontana.

(N&FD Photo)

(NSFD Photo)

Exterior attack in progress; note fire involvement in attic.

(M&FD PA photo)

Front of hotel.

Aerial view of hotel after the fire. Note destruction of the roof and proximity of the Prince Michael Hotel.

(WSFD Photo)

Front of hotel, lobby desk is located to far right along wall.

Front of hotel, looking toward the Prince Michael Hotel.

Lobby, looking up through roof facing Collins Street.

First floor corridor, looking to rear of building from lobby. Fire door in foreground was closed throughout much of the fire.

Typical floor construction, shown from ceiling of first floor room.

Lobby, looking up pipe chase located adjacent to elevator shaft.

Outline of victim in smoke stain, Room 205. Victim was reportedly found carrying a flashlight (still on) and personal belongings in a purse.

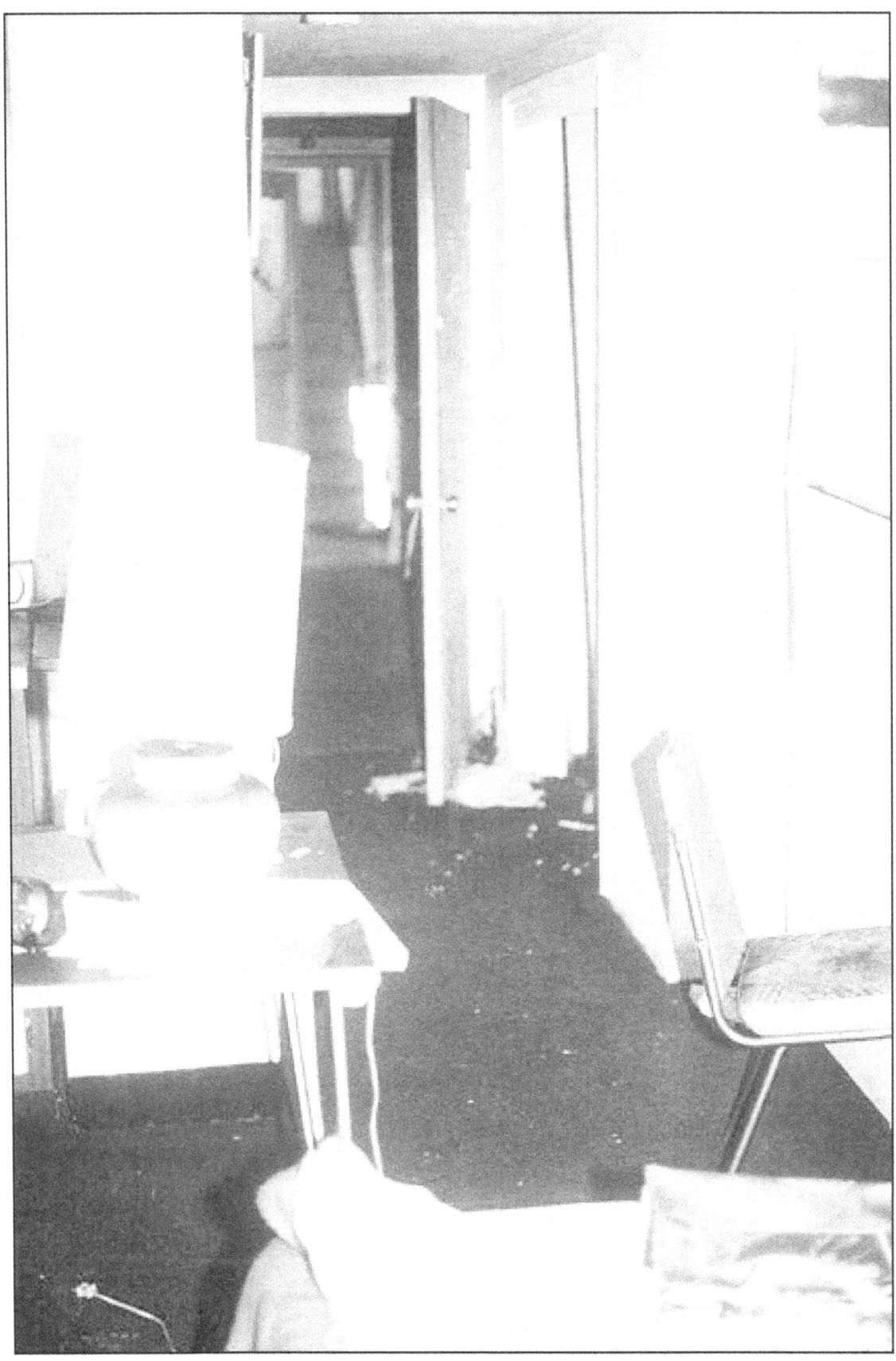

View of Room 205, fire exit is visible across corridor.

Second floor, looking across to Room 301 from front stairway, area of origin is at base of photo.

View from second floor, front stairway looking straight across building.
Lobby is at bottom of photo.

Manual fire alarm station and bell, third floor, at rear stair. Note extensive darkening soot.

Remains of fire alarm control panel.

U.S. Government Printing Office: 2000-722205494278